Conservation of ... what?

Many times in physics you will make use of the conservation idea. This is a simple idea but turns out to be very powerful. In everyday language, it comes down to
'Whatever you start out with, you must end up with.'
Whatever changes may have happened along the way and whatever 'disguises' might be involved, the conservation idea remains true.

Newton's cradle is a good example. Once you have provided the initial energy to set the balls in motion, then the total energy in the system remains the same whenever you calculate its value. Some energy will appear as kinetic energy, some as potential energy, and some as heat energy or sound energy, but the total will always be the same.

Therefore if you ever add up the total and some energy is 'missing', then you know you haven't thought about *all* the energy transfers going on and it prompts you to look a little more closely, perhaps discovering something you hadn't known before. Geiger and Marsden made a critical discovery about the structure of atoms when they noticed that the numbers in their experiment didn't quite add up.

There are other conservation rules. For example, we use ideas about
almost any change, the total number of atoms at the end must be th
one time this was thought to be always true. They might be rearranged into complicated new patterns or new states, but all the particles must be there. For example, think about what a tree is made from. Combining atoms, making sure that we account for all of them, allows us to 'build' the tree. Using just carbon dioxide from the air, water from the soil and traces of other simple chemicals, the atoms combine to form the largest living things ever seen on the Earth.

But physics is also a developing science – it is not finished. During the development of nuclear physics in the 20th century, careful experiments showed that in some situations mass was *not* conserved and neither was the energy. Atoms can even appear and disappear.
Mass seemed to appear from nowhere and energy disappeared altogether.
A deeper truth was revealing itself and now, in some situations, we need to
make use of the full conservation rule, the *conservation of mass-energy*.
A rule that involves the most famous physics equation of them all:

$$E = mc^2$$

Albert Einstein (1879–1955).
Although famous for his papers on special and general relativity, he won the 1921 Nobel Prize for Physics for his work on the photoelectric effect.

William Collins' dream of knowledge for all began with the publication of his first book in 1819. A self-educated mill worker, he not only enriched millions of lives, but also founded a flourishing publishing house.

Today, staying true to this spirit, Collins books are packed with inspiration, innovation and practical expertise. They place you at the centre of a world of possibility and give you exactly what you need to explore it.

Collins

DO MORE

Published by Collins
An imprint of HarperCollinsPublishers
77 – 85 Fulham Palace Road,
Hammersmith,
London W6 8JB

Browse the complete Collins Education catalogue at
www.collinseducation.com

© HarperCollinsPublishers Limited 2006

10 9 8 7 6 5

ISBN-10: 0-00-775543-0
ISBN-13: 978-0-00-775543-1

British Library Cataloguing in Publication Data
A Catalogue record for this publication is available from the
British Library

Cover design by White-Card, London
New text by Derek Wade and Tim Jolly
Edited by Kay Macmullan
Text page design by Sally Boothroyd
New artwork by Jerry Fowler
Printed and bound by Printing Express, Hong Kong

Acknowledgements:
The Authors and Publishers are grateful to the following for
permission to reproduce copyright material:

University of Cambridge International Examinations: pp 50 – 53,
82 – 83, 104 – 105, 156 – 159, 170 – 17, 1841

University of Cambridge International Examinations bears no
responsibility for the example answers to questions taken from its
past question papers which are contained in this publication.

Photographs
© Jupiterimages Corporation (c) 2006 9, 11, 18, 20, 21(L), 21 (R), 22,
25, 26, 36 (L), 36 (R), 40, 44 (L), 44 (R), 47, 49, 56 (L), 56 (R), 58, 59,
62 (L), 62 (R), 66, 67, 75, 86, 89, 91, 94, 98(T), 98(B), 101, 114, 134
(T), 134 (B), 147, 149, 164, 167, 168 (T), 168 (B), 202-205; Pbase 112;
Science Photo Library 97; David Vincent 4 (L), 4 (R), 23, 45 (T), 45 (B),
109, 110 (L), 110 (R), 150 (A), 150 (B), 150 (C), 150 (D)

Cover illustration: Newton's Cradle – Tim Ridley ©, Dorling Kindersley

Inside Front Cover spread: Isaac Newton – Photo Researchers /
Science Photo Library; Newton's Cradle – Martyn F. Chillmaid /
Science Photo Library; Bouncing Ball – Adam Hart-Davis / Science
Photo Library; Albert Einstein – US Library of Congress / Science
Photo Library

Section spreads: pp6/7 Petronas Towers – © Jose Fuste Raga/Corbis;
Taipei 101 tower – © Simon Kwong/Reuters/Corbis; pp54/55
Collection of Succulents – Andrew McRobb/DK Images © Dorling
Kindersley; Molecular structure of water vapour – Clive Freeman
Biosym Technologies/Science Photo Library; pp84/85 Envisat-1 satellite
in orbit – European Space Agency/Science Photo Library; Singapore
and surroundings – NASA/Science Photo Library; pp106/107 Arctic
Tern – © Arthur Morris/Birds As Art/NHMPL; Earth's Magnetic Field
– Gary Hincks/Science Photo Library; pp 160 Gamma Scan of Skull &
Spine – Philippe Plailly/Science Photo Library

Every effort has been made to contact the holders of copyright
material, but if any have been inadvertently overlooked, the Publishers
will be pleased to make the necessary arrangements at the first
opportunity.

Cambridge IGCSE
PHYSICS

by Malcolm Bradley & Chris Sunley
Consultant Editor: Derek Wade

Collins

GETTING THE BEST FROM THE BOOK

Welcome to *Cambridge IGCSE Physics*. This textbook and the accompanying CD-ROM have been designed to help you understand all of the requirements needed to succeed in the Cambridge IGCSE Physics course. Just as there are five sections in the Cambridge syllabus so there are five sections in the textbook: General physics; Thermal physics; Properties of waves, including light and sound; Electricity and magnetism and Atomic physics. Each section in the book covers the essential knowledge and skills you need. The textbook also has some very useful features that have been designed to help you understand all the aspects of Physics that you will need to know for this specification.

Coverage of each topic is linked closely to the Cambridge specification so that you build a powerful knowledge-base with which to succeed in the examination.

Photographic images help you to visualise the information you need.

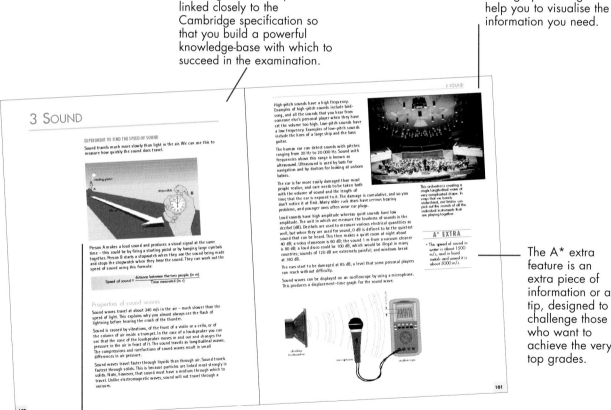

The A* extra feature is an extra piece of information or a tip, designed to challenge those who want to achieve the very top grades.

Clear illustrations help you to understand physical processes and structures.

Cambridge IGCSE Physics is assessed in the following manner:

Paper 1 – Compulsory multiple choice paper – worth 30% of the marks

AND EITHER
Paper 2 – The Core Curriculum (Grades C to G) – worth 50% of the marks
OR
Paper 3 – The Extended Curriculum (Grades A* to G) – worth 50% of the marks

AND EITHER
Paper 4 – Coursework (common to both tiers) – worth 20% of the marks
OR
Paper 5 – Practical Test (common to both tiers) – worth 20% of the marks
OR
Paper 6 – Alternative to Practical (common to both tiers) – worth 20% of the marks.

Collins *Cambridge IGCSE Physics* covers all of the topics and skills you will need to achieve success, whichever assessment pathway you are entered for.

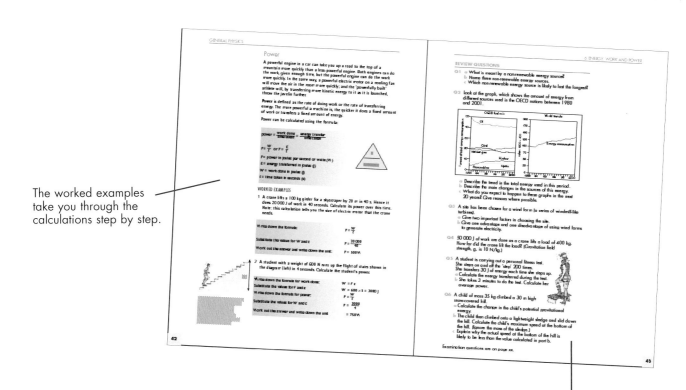

The worked examples take you through the calculations step by step.

Banks of questions appear after every topic to test your understanding of the work you've just covered. All the answers can be found using the information in the text.

GETTING THE BEST FROM THE CD-ROM

IGCSE Physics CD-ROM

To help you through the course we have added this unique CD-ROM, which you may be able to use in class or as part of your private study. To allow you to understand the subject as you progress through the course we have added the following features to the CD-ROM:

LIST OF VIDEOS ON THE PHYSICS CD ROM

1. mass + weight + momentum
2. Hooke's Law
3. Constant velocity
4. $F = ma$
5. Electrostatics
6. The volume and density of an irregular shape
7. Terminal velocity
8. Pressure
9. Circuits
10. Simple parallel circuit
11. Iron filings/compass
12. Making an electromagnet
13. Making a motor/Faraday motor
14. Heat: thermal expansion and contraction
15. Types of heat transfer
16. The spectrum
17. Ray boxes/reflection with mirrors
18. Refraction, reflection, diffraction
19. Radioactivity

VIDEO CLIPS

Physics is a practical subject and to reinforce your studies the CD-ROM includes 19 short films that demonstrate experiments and practical work. Each of the films covers an area that you need to know well for the practical section of the course. The information in the films will also help you with the knowledge required in other parts of the specification. You can use the films to help you understand the topic you are currently studying or perhaps come back to them when you want to revise for the exam. Each of the films has sound too, so if you are watching it in a library or quiet study area you may need headphones.

QUESTION BANK

'Practice makes perfect', the saying goes, and we have included a large bank of questions related to the Physics syllabus to help you understand the topics you will be studying.

As with the films, your teacher may use this in class or you may want to try the questions in your private study sessions.

These questions will reinforce the knowledge you have gained in the classroom and through using the textbook and could also be used when you are revising for your examinations. Don't try to do all the questions at once though; the most effective way to use this feature is by trying some of the questions every now and then to test yourself. In this way you will know where you need to do a little more work. The questions are not full 'exam-type' questions that you will be set by your IGCSE examiners. Some of the questions test underlying principles that are not specifically mentioned in your specification.

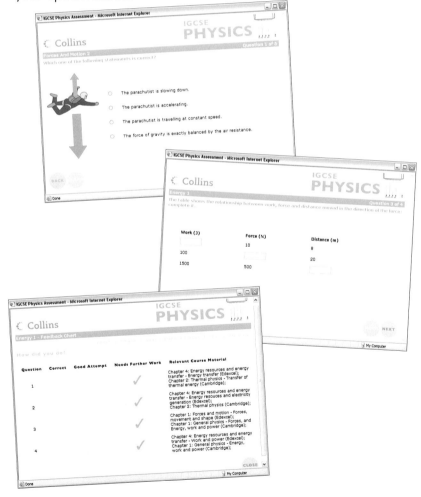

Good luck with your IGCSE Physics studies. This book and the CD-ROM provide you with stimulating, interesting and motivating learning resources that we are sure will help you succeed in your Physics course.

OPERATING SYSTEMS REQUIRED AND SET-UP INSTRUCTIONS.

Mac System requirements
500 MHz PowerPC G3 and later
Mac OS X 10.1.x and above
128MB RAM
Microsoft Internet Explorer 5.2, Firefox 1.x, Mozilla 1.x, Netscape 7.x and above, Opera 6, or Safari 1.x and above (Mac OS X 10.2.x only)
325 MB of free hard disc space.

To run the program from the CD
1 Insert the CD into the drive
2 When the CD icon appears on the desktop, double-click it
3 Double-click Collins IGCSE Physics.html

To install the program to run from the hard drive
1 Insert the CD into the drive
2 When the CD icon appears on the desktop, double-click it to open a finder window
3 Drag Collins IGCSE Physics.html to the desktop
4 Drag Collins IGCSE Physics Content to the desktop.

PC System requirements
450 MHz Intel Pentium II processor (or equivalent) and later
Windows 98/ME/NT/2000/XP
128MB RAM
Microsoft Internet Explorer 5.5, Firefox 1.x, Mozilla 1.x, Netscape 7.x and above, Opera 7.11 and above
325 MB of hard disc space

To run the program from the CD
1 Insert the CD into the drive
2 Double-click on the CD-ROM drive icon inside My Computer
3 Double-click on Collins IGCSE Physics.html

To install the program to run from the hard drive
1 Insert the IGCSE Physics disc into your CD-ROM drive
2 Double-click on the CD-ROM drive icon inside My Computer
3 Double-click on the SETUP.EXE
4 Follow onscreen instructions. These include instructions concerning the Macromedia Flash Player included with and required by the program.
5 When the installation is complete, remove the CD from the drive.

For free technical support, call our helpline on: Tel.: + 44 141 306 3322 or send an email to: it.helpdesk@harpercollins.co.uk.

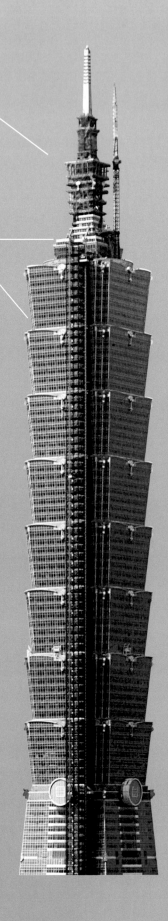

Completed in 2003, Taipei 101 (because it has 101 floors) holds the 'world's tallest' in several categories, such as the highest occupied floor and the world's fastest elevators, rising from the basement to the 89th floor in 39 seconds

101st floor

The 87th floor holds a 900-ton tuned-mass damper to counteract earthquakes

Reaching for the sky

Have you ever wondered just how much physics is involved in building a tall building?

For a start, think about some of the forces. Gravity is trying to pull your building down so you must use materials that have enough strength to balance this, and what you use to join them together must be just as strong. But don't forget the wind. The wind is trying to turn your building over sideways; your design will have to withstand considerable turning effects. And what about inside the building? You need to create lots of space so the building is useful. You will need to know about forces and how to control them.

People need to get about inside the building too. You will need some elevators, but how long will these take to reach the top? People won't want to spend an hour reaching their floor. So how fast can an elevator travel and how quickly can it accelerate without making the people inside feel ill? You are going to need to know about motion.

GENERAL PHYSICS

The Petronas Towers in Kuala Lumpur.
Completed in 1998

88th floor

'Sky bridge' connects the two buildings
at the 41st and 42nd floors

The world's deepest foundations,
reaching 120 m down to bedrock

1 LENGTH AND TIME

This train runs for 30 km between Shanghai and Pudong Airport, and completes the journey in 7 minutes, reaching a top speed of 430 km/h. The train uses magnets to hover 10 mm above the track. The track must be placed within a few mm of the planned route, and this requires great accuracy in all measurements.

Making measurements is very important in physics. Without numerical measurements physicists would have to rely on descriptions, which could lead to inaccurate comparisons. Imagine trying to build a house if the only instructions were 'big' and 'small'.

When making measurements, physicists use different instruments, such as rules to measure lengths, measuring cylinders to measure volume and clocks to measure time.

A physicist will always take care to make the measurements as accurate as possible. If she is using a rule she will place the rule along the object to be measured, and will read off the scale the positions of the beginning and the end of the object. The length is, of course, the difference between these two readings. If the rule is nearer to her eye than the object being measured, the reading will appear to change as she moves her eye. The correct reading is obtained when her eye is directly above the point being measured.

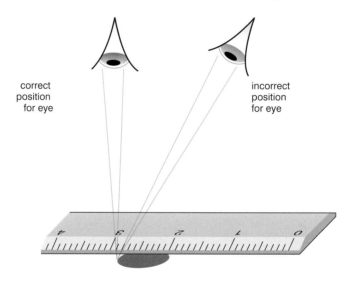

To improve accuracy further, she may take several readings, and use the average of these readings as a better result.

To use a measuring cylinder, she will firstly make sure that the cylinder is standing on a level table. Then she will make sure that her eye is at the same level as the liquid inside the cylinder. The surface of most liquids will bend up or down near the walls of the measuring cylinder. This bent shape is known as a meniscus. However, most of the surface is flat, and measurements are made to this flat surface.

Warning: Some measuring cylinders have unusual scales, and one division may represent an unexpected quantity, perhaps 2 cm^3 or 0.5 cm^3. Check carefully.

In this book, volumes will usually be measured in cm^3 (or perhaps in m^3). In other places, such as on some measuring cylinders, you will see the millilitre.

A volume of 1 mL is the same as a volume of 1 cm^3.
1000 cm^3 = 1000 mL = 1 L

For measuring large volumes we also use the cubic metre.
$1\ m^3 = 1000\ L = 1000\ 000\ cm^3$

You may need to use specialised measuring equipment. For example, the micrometer is used to measure small distances, such as the diameter of a piece of wire.

The micrometer is designed so that the gap between the jaws changes by 0.5 mm for every complete turn of the thimble. By measuring the exact position of the thimble when an object is being held, the thickness of the object can be measured very accurately. A physicist will always check that the jaws of the micrometer are clean, and will then check that the reading is 0 mm when the jaws are closed gently. Most micrometers allow the zero reading to be reset, but this may need to be done by a trained person. Most micrometers also have a special ratchet fitted onto the end of the thimble, which slips and emits a clicking sound when sufficient force has been applied, but extra care should be taken if the micrometer does not have one of these.

To measure the thickness of an object, open the jaws of the micrometer and close them gently onto the object in question. A scale on the barrel will show by how many complete turns the jaws have been opened, with every two turns indicating another millimetre. The scale round the edge of the thimble is calibrated from 0 to 50. So a reading of 40 indicates that a further 0.40 mm must be added to the thickness. But be careful, a reading of 5, say, indicates that only 0.05 mm is to be added.

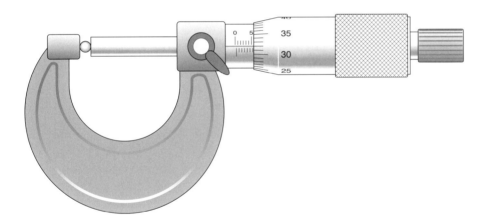

The marks along the top of the line along the barrel show that the jaws have been opened to 5 mm, and the fact that an additional mark has become visible below the line shows that they are opened beyond 5.5 mm. Therefore you know that the answer must be between 5.5 mm and 6.0 mm. Next you look at the scale on the thimble. The reading of 32 shows that you must add 0.32 mm to the reading. So the final answer is 5.82 mm.

Times are measured by using a stopwatch or stopclock.

Hand-operated stopwatches have an accuracy that is limited by the delay between your eye seeing the moment to start, your brain issuing the command to start the watch and your finger pressing the start button. The total delay is typically around 0.2 s. This delay is known as your 'reaction time', and it increases the danger of some tasks, such as driving a car.

For applications where accuracy is critical, such as in athletics, the clock has to be started and stopped automatically by the athlete breaking a light beam that shines across the track.

If you are measuring the time of an oscillation, such as the swing of a pendulum, it is very easy to improve the accuracy of the measurement by timing a number of swings, perhaps 10 or 20.

The reaction time of the person using the stopwatch will affect the accuracy of the timing.

It is important to count correctly. Let the swing go, count zero and start the stopwatch as the pendulum crosses a mark at the bottom of the swing (we call this the fiducial mark). The next time the pendulum crosses the fiducial mark going in the same direction count one, and so on. In this way the count will be correct.

After measuring the time for 20 swings, say, divide the total time by 20 to give the period of one oscillation of the pendulum.

REVIEW QUESTIONS

Q1 Rules that are 30 cm long are often made of wood or plastic that is thicker in the middle, and thinner along the edges where the scale is printed. Explain why the user is less likely to make an error if the rule is thinner at the edge, and suggest reasons why the rule is thicker in the middle.

Q2 A plastic measuring cylinder is filled with water to the 100 cm^3 mark, and a student measures the column of water in the cylinder and finds that it is 20 cm high.
  a The student pours 10 cm^3 of the water out of the cylinder. How high will the column of water be now?
  b The student then refills the cylinder back to the 100 cm^3 mark by holding it under a dripping tap. She finds that it takes 180 drops of water. What is the volume of one of these drops?
  c What is the cross-sectional area of the cylinder? Hint: The volume of a cylinder is given by the equation: volume = cross-sectional area × length.
  d So from answer (c) what is the internal diameter of the tube used to make the measuring cylinder?

Q3 A student tries to measure the period of a pendulum that is already swinging left and right. At the moment when the pendulum is fully to the left, she counts 'One' and starts a stopwatch. She counts successive swings each time that the pendulum returns to the left. When she counts 'Ten' she stops the stopwatch, and sees that it reads 12.0 s.
  a What was her mistake?
  b What is the period of swing of this pendulum?
  c In this particular experiment, explain the likely effect of her reaction time on her answer.

More questions on the CD ROM

Examination questions are on page 50.

2 SPEED, VELOCITY AND ACCELERATION

We have all been in a car travelling at 90 kilometres per hour. This, of course, means that the car (if it kept travelling at this speed for one hour) would travel 90 km. During one second of its journey this car travels 25 metres, so its speed can also be described as 25 metres per second. Scientists prefer to measure time in seconds, and distance in metres. So they prefer to measure speed in metres per second, often written as m/s.

The **speed** of an object can be calculated using the following formula:

$$\text{speed} = \frac{\text{distance}}{\text{time}}$$

$$v = \frac{s}{t}$$

v = speed in m/s
s = distance in m
t = time in s

Most objects speed up and slow down as they travel. An object's 'average speed' can be calculated by dividing the total distance travelled by the total time taken.

WORKED EXAMPLES

1 Calculate the average speed of a motor car that travels 500 m in 20 seconds.

Write down the formula: $v = \frac{s}{t}$

Substitute the values for s and t: $v = \frac{500}{20}$

Work out the answer and write down the units: $v = 25$ m/s

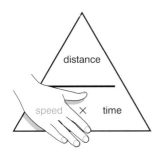

Cover speed to find that
$$\text{speed} = \frac{\text{distance}}{\text{time}}$$

2 A horse canters at an average speed of 5 m/s for 2 minutes. Calculate the distance it travels.

Write down the formula in terms of s: $s = v \times t$

Substitute the values for v and t: $s = 5 \times 2 \times 60$

Work out the answer and write down the units: $s = 600$ m

Cover distance to find that

distance = speed × time

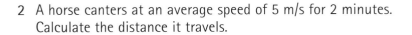

Each car's average speed can be calculated by dividing the distance it has travelled by the time it has taken.

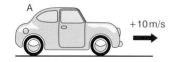

Both cars have the same speed. Car A has a velocity of +10 m/s, car B has a velocity of –10 m/s.

A distance–time graph for a bicycle travelling down a hill. The graph slopes when the bicycle is moving. The slope gets steeper when the bicycle goes faster. The slope is straight (has a constant gradient) when the bicycle's speed is constant. The line is horizontal when the bicycle is at rest.

ARE SPEED AND VELOCITY THE SAME?

We often want to know the direction in which an object is travelling. For example, when a space rocket is launched, it is likely to reach a speed of 1000 km/h after about 30 seconds. However, it is extremely important to know whether this speed is upwards or downwards. You want to know the speed *and* the direction of the rocket. The velocity of an object is one piece of information, but it consists of two parts: the speed and the direction. In this case, the velocity of the rocket is 278 m/s (its speed) upwards (its direction).

A velocity can have a minus sign. This tells you that the object is travelling in the opposite direction. So a velocity of -278 m/s upwards is actually a velocity of 278 m/s downwards.

USING GRAPHS TO STUDY MOTION

Journeys can be summarised using **graphs**. The simplest type is a **distance–time graph** where the distance travelled is plotted against the time of the journey.

At the beginning of an experiment, time is usually given as 0 s, and the position of the object 0 m. If the object is not moving, then time increases, but distance does not. This gives a horizontal line. If the object is travelling at a steady speed, then both time and distance increase steadily, which gives a straight line. If the speed is varying, then the line will not be straight.

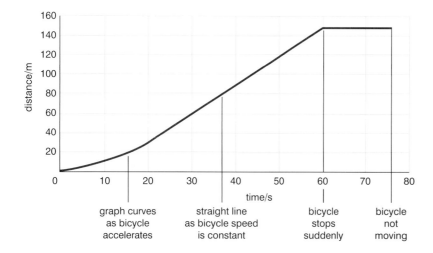

graph curves as bicycle accelerates

straight line as bicycle speed is constant

bicycle stops suddenly

bicycle not moving

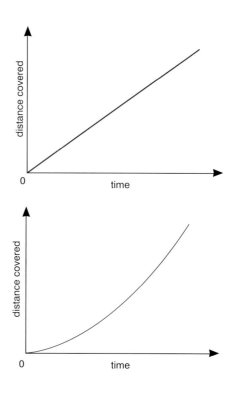

Steady speed is shown by a straight line. Steady acceleration is shown by a smooth curve.

WHAT IS ACCELERATION?

If the speedometer of a car displays 50 km/h, and then a few seconds later it displays 70 km/h, then the car is accelerating. If the car is slowing down, this is called negative acceleration, or deceleration. Acceleration is a change in speed or velocity.

Let us imagine that the car is initially travelling at 15 m/s, and that one second later it has reached 17 m/s, and that its speed increases by 2 m/s each second after that. Each second its speed increases by 2 metres per second. We can say that its speed is increasing at '2 metres per second *per second*'. This can be written, much more conveniently, as an acceleration of 2 m/s^2.

Our planet Earth attracts all objects towards its centre with the force of gravity. The strength of the force decreases slowly with distance from the surface of the Earth, but for objects within a few km of the surface, all objects that are falling freely will have the same constant acceleration of just under 10 m/s^2. If a coconut falls from a tree, then after 1 s it will be falling at 10 m/s (though it will only have travelled 5 m because, of course, it started with zero velocity). After 2 s it will be falling at 20 m/s, if it does not hit the ground first.

How much an object's **speed or velocity changes** in a certain time is its **acceleration**. Acceleration can be calculated using the following formula:

$$\text{acceleration} = \frac{\text{change in velocity}}{\text{time taken}}$$

$$a = \frac{(v - u)}{t}$$

a = acceleration

v = final velocity in m/s

u = starting velocity in m/s

t = time in s

A* EXTRA

- A negative acceleration shows that the object is slowing down.

WORKED EXAMPLE

Calculate the acceleration of a car that travels from 0 m/s to 28 m/s in 10 seconds.

Write down the formula:

$$a = \frac{(v - u)}{t}$$

Substitute the values for v, u and t:

$$a = \frac{(28 - 0)}{10}$$

Work out the answer and write down the units:

$$a = 2.8 \text{ m/s}^2$$

A **velocity–time graph** provides information on speed or velocity, acceleration and distance travelled.

Steady velocity is shown by a horizontal line. Steady acceleration is shown by a line sloping up.

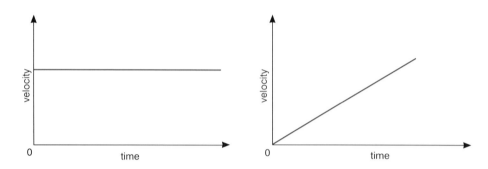

In the graph above left, the object is already moving when the graph begins. If the object starts with a velocity of zero, then the line starts from the origin as shown in the graph above right.

Note that the object may not move to begin with. In this case the line will start by going along the *x*-axis, showing that the velocity stays at zero for a while.

MEASURING AREA UNDER A GRAPH

The area under a velocity-time graph gives you the distance travelled, because distance = velocity × time. Always make sure the units are consistent, so if the velocity is in km/h, you must use time in hours too.

The graph below shows how the velocity of a car varies as it travels between two sets of traffic lights. The graph can be divided into three regions.

A velocity-time graph for a car travelling between two sets of traffic lights.

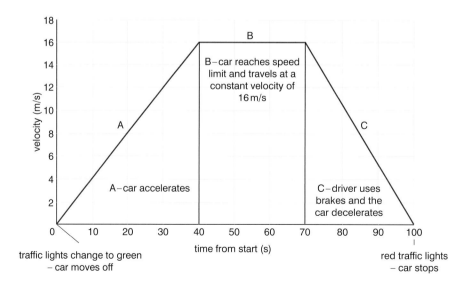

In region A, the car has constant **acceleration** (the line has a constant positive gradient). The distance travelled by the car can be calculated:

average velocity $= \dfrac{(16 + 0)}{2} = 8$ m/s

time = 40 s

so distance $= v \times t = 8 \times 40 = 320$ m

This can also be calculated from the area under the line
($\frac{1}{2}$ base $\times$ height $= \frac{1}{2} \times 40 \times 16 = 320$ m).

In region B, the car is travelling at a **constant velocity** (the line has a gradient of zero). The distance travelled by the car can be calculated:

velocity = 16 m/s

time = 30 s

so, distance $= v \times t = 16 \times 30 = 480$ m

This can also be calculated from the area under the line
(base $\times$ height $= 30 \times 16 = 480$ m).

In region C, the car is **decelerating at a constant rate** (the line has a constant negative gradient). The distance travelled by the car can be calculated:

average velocity $= \dfrac{(16 + 0)}{2} = 8$ m/s

time = 30 s

so, distance $= v \times t = 8 \times 30 = 240$ m

This can also be calculated from the area under the line
($\frac{1}{2}$ base $\times$ height $= \frac{1}{2} \times 30 \times 16 = 240$ m).

In the above example, the acceleration and deceleration were constant, and the lines in regions A and C were straight. This is very often not the case. You will probably have noticed that a car will have a larger acceleration when it is travelling at 30 km/h than when it is travelling at 120 km/h.

A man-carrying space rocket does exactly the opposite, and if you watch one being launched you can see that it has a small acceleration. As it is burning several tonnes of fuel per second, it quickly becomes less massive and starts to have a larger acceleration.

REVIEW QUESTIONS

Q1 An aeroplane takes off from an airfield and travels north for 20 minutes and the pilot finds herself 200 km north of her starting point. She finds that she has gone too far, and she travels south for 10 minutes at 300 km/h to reach the airfield that is her destination.
  a What is the plane's average speed for the first part of its journey?
  b What is the plane's average speed for the whole journey?
  c How far apart are the two airfields?

Q2 The graph shows a distance–time graph for a journey.
  a What does the graph tell us about the speed of the car between 20 and 60 seconds?
  b How far did the car travel between 20 and 60 seconds?
  c Calculate the speed of the car between 20 and 60 seconds.
  d What happened to the car between 80 and 100 seconds?

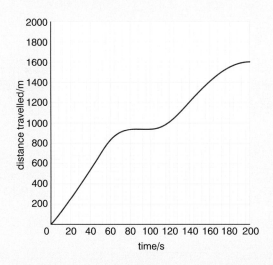

Q3 Look at the velocity–time graph for a toy tractor.
  a Calculate the acceleration of the tractor from A to B.
  b Calculate the total distance travelled by the tractor from A to C.

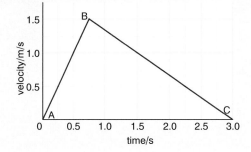

More questions on the CD ROM

Examination questions are on page 50.

3 MASS AND WEIGHT

Scientists use the words 'mass' and 'weight' with special meanings. By the 'mass' of an object we mean how much material is present in it.

Weight is the force on the object due to gravity. It is measured in **newtons**. The weight of an object depends on its **mass** and **gravitational field strength**. Any mass near the Earth has weight due to the Earth's gravitational pull.

Weight is calculated using the equation:

Weight (W) = mass (m) × gravitational field strength (g)

Scientists often use the word 'field'. We say that there is a 'gravitational field' around the Earth, and that any object that enters this field will be attracted to the Earth.

The value of the gravitational field strength on Earth is 9.8 N/kg, though we usually round it up to 10 N/kg to make the calculations easier. A gravitational force of 10 N acts on an object of mass 1 kg on the Earth's surface.

Note that gravity does not stop suddenly as you leave the Earth. Satellites go round the Earth and do not escape, because the Earth is still pulling them, even if less strongly than before the satellites were launched. The Earth is even pulling the Moon gently, and this is why it orbits the Earth once per month. And the Earth goes round the Sun because the Sun's gravity is pulling the Earth.

If you stand on the Moon you will feel the gravity of the Moon pulling you down. Your mass will be the same as on Earth, but your weight will be less. This is because the gravitational field strength on the Moon is about one-sixth of that on the Earth, and so the force of attraction of an object to the Moon is about one-sixth of that on the Earth. The gravitational field strength on the Moon is 1.6 N/kg.

Earth

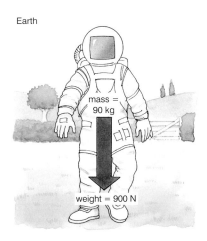

mass = 90 kg

weight = 900 N

Moon

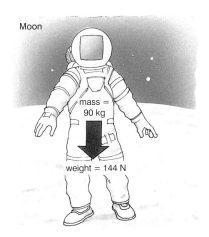

mass = 90 kg

weight = 144 N

Though your mass remains the same, your weight is greater on Earth than it would be on the Moon.

If two astronauts played American football, it would just as difficult to halt a rush by one of them on the Moon as it would be on the Earth, and any collision between them would hurt just as much. The reason is that it is the mass of an object that resists any change in the motion of the object, and the mass of each astronaut is the same in both places.

It is harder to get a massive object moving, and it is harder to stop it once it is moving.

A supertanker laden with oil and travelling at 18 km/hr will take over 12 km to stop. A speedboat travelling at the same speed will take less than 100 m. The difference is due to the mass of the tanker.

HOW DO YOU WEIGH SOMETHING?

The balance is level when the forces pulling down both sides are the same. In the balance shown, the forces of 10 N and 20 N on the one side balance the force of 30 N on the other side. The balance compares the weight of the objects on each side. If the balance is on the surface of the Earth, then the masses of these objects are 1 kg and 2 kg on one side, and 3 kg on the other. So the balance also allows you to compare masses.

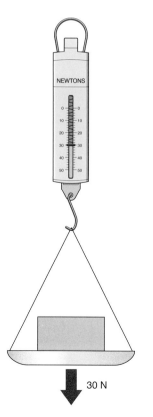

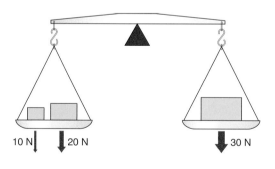

Note that this type of balance would also work on the Moon.

THE SPRING BALANCE

A spring balance can also be used for weighing things, but works in a different way.

The top of the spring is hung from a hook, and the spring is stretched by the weight of the pan attached to its lower end. The scale can then be adjusted so that the pointer is aligned with the 'zero' mark.

When a known mass is placed in the pan, the spring stretches further due to the extra weight, and the new pointer position can be marked. In the spring balance shown, the pointer should be at the 30 N mark if the scale is set correctly. If this balance were moved to the Moon, the weight would be less, and the spring would not stretch so far. In fact the pointer would indicate a weight of 4.8 N.

So the spring balance measures the weight of the object in newtons. For non-scientific use, these balances are often given a scale that indicates the mass of the object in kg, without the need for any calculations. This scale gives the correct mass on the surface of the Earth, but would definitely not give the correct mass if the spring balance were moved to the Moon.

See page 28 for more about springs.

REVIEW QUESTIONS

Note that the gravitational field strength on the surface of Mars is 3.8 N/kg.

Q1 A teenage astronaut has a mass of 60 kg when she gets into her spacecraft on Earth.
  a What is her weight on Earth?

Parts b–e refer to the situation on the surface of Mars.
  b What is her mass now?
  c What is her weight now?
  d If she stands in one pan of a large balance, what masses would be needed in the other pan to balance her?
  e If she stands on bathroom scales (which are a type of spring balance) what would be the reading in Newtons?

Q2 The height that you can jump depends inversely on the gravitational field strength. So if the field strength doubles, the height halves. If the Olympic Games were held on Mars in a large dome to provide air to breathe, what would happen to the records for:
  a weight lifting (weight in N)
  b high jump (height)
  c pole vault (height)
  d throwing the javelin (distance)
  e the 100 m race (time)?
  In every case, state whether the record is likely to increase, stay similar or decrease, and explain your choice.

More questions on the CD ROM

Examination questions are on page 50.

4 DENSITY

Gold is one of the densest metals. A block the size of a 1 L carton of milk would have a mass of almost 20 kg and would be very hard to pick up.

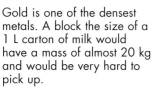

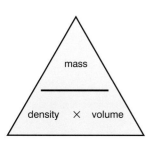

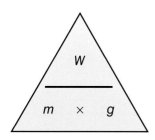

We must have all noticed that the weight of objects can vary greatly. A plastic teaspoon weighs less than a metal one, and a gold ring weighs twice as much as a silver one, even if the objects are exactly the same size.

The density of a material is a measure of how 'squashed up' it is, and a dense object contains more mass than a light object of the same size.

Density is calculated using this formula:

$$d = \frac{m}{V}$$

m = mass in g or kg

V = volume in cm³ or m³

d = density in g/cm³ or kg/m³

Note that in this equation you must use g and cm throughout or you must use kg and m. And note that if you measure the weight in N you must convert it into g or kg.

THE DENSITY OF A REGULARLY SHAPED OBJECT

WORKED EXAMPLE

The brick has dimensions 20 cm x 9 cm x 6.5 cm.
Weight of brick = 22.2 N
What is the density of the brick?

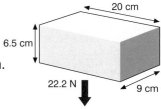

Mass of brick, m	$= \dfrac{W}{g}$
	$= \dfrac{22.2}{10}$ kg
	$= 2.22$ kg
	$= 2220$ g
(Remember that 1 kg = 1000 g)	
Volume of brick, V	$= 20 \times 9 \times 6.5$ cm³
	$= 1170$ cm³
Density of brick, d	$= \dfrac{mass}{volume}$
	$= \dfrac{2220}{1170}$ g/cm³
	$= 1.90$ g/cm³

Note that the density of water is 1.0 g/cm³, and the rule is that an object of greater density will sink in a liquid of lower density. So, perhaps not surprisingly, the brick will sink in water. But will it sink in mercury? See the table right.

Some useful densities

	Density in g/cm^3	Density in kg/m^3
Vacuum	0	0
Helium gas	0.00017	0.17
Air	0.00124	1.24
Oil (Petroleum)	0.88	880
Water	1.0	1000
Seawater	1.03	1030
Plastic	0.9 – 1.6	900 – 1600
Wood	0.5 – 1.3	500 – 1300
Magnesium	1.74	1740
Aluminium	2.7	2700
Titanium	4.5	4500
Steel	7.8	7800
Mercury (liquid)	13.6	13600
Silver	10.5	10500
Gold	19.3	19300

WHY DO MATERIALS HAVE DIFFERENT DENSITIES?

If you look inside a block of gold and inside a block of aluminium with a modern electron microscope, you will see that the atoms are almost exactly the same size (the gold atoms are just a little bit bigger). As we will see later, most of an atom is actually empty space, and the mass of an atom is concentrated in the nucleus, which is far smaller than the atom. So the extra density of the gold is due to the fact that the nuclei of the gold atoms are far more massive than the nuclei of the aluminium ones.

Many materials have a lower density because they contain large air bubbles or other voids inside them. Bread has a lower density than most cakes; and expanded polystyrene cups have a lower density than other cups.

A bag of popcorn has a far lower density than the same bag filled with corn that has not been popped.

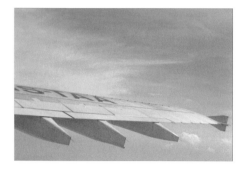

The choice of materials used to make an aircraft is critical in making it as light as possible and thus reducing fuel consumption.

An aeroplane is another example of a lower density. Although aeroplanes are made of aluminium and other metals of lower density, there is no way that an aeroplane could fly if it was made of solid aluminium. In fact, because most of the inside of an aeroplane consists of air, the average densities of all aeroplanes are sufficiently low that, in the event of a forced landing on water, they can easily float for long enough for everyone to escape.

WORKED EXAMPLE

What is the mass of a block of expanded polystyrene that is 1 m long, 0.5 m wide and 0.3 m high? The density of this sample of expanded polystyrene is 8 kg/m^3.

Volume of block, V $= 1.0 \times 0.5 \times 0.3$ m^3

                                      $= 0.15$ m^3

Write down the formula: $m = d \times V$

Substitute the values for d and V: $m = 8 \times 0.15$ kg

Work out the answer and write down the units: $m = 1.2$ kg

MEASURING THE DENSITY OF AN IRREGULAR OBJECT

This method only works if the object is denser than the liquid used so that it sinks. It does not work if the object absorbs the liquid, nor if it is damaged by the liquid.

Step 1

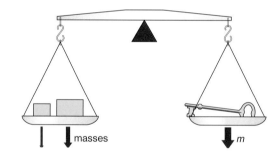

masses m

Use a balance to weigh the object in question and so find its mass, m.

Step 2

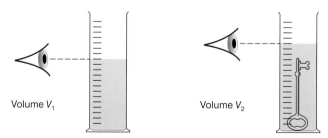

Volume V_1 Volume V_2

Choose a measuring cylinder that will accept the object. A narrower cylinder will give a more accurate answer than a wider one. Add liquid to the cylinder to fill it to a deep enough level so that the object will be completely submerged, and then measure the volume of liquid V_1. The exact amount of liquid that you use is not at all critical. Water is often the liquid used.

Step 3

Lower the object into the liquid (without splashing) and measure the new reading V_2. This is the volume of the object and the liquid. The volume of the object is $V_2 - V_1$.

From the mass and the volume you can calculate the density of the object.

WORKED EXAMPLE

A small metal statue is measured to have a mass of 90 g.
A measuring cylinder is filled with water to the 82 cm^3 mark. The statue is lowered into the measuring cylinder and the water rises to the 91 cm^3 mark. What is the statue made of?

Volume of the statue   = 91 – 82 cm^3

                                    = 9 cm^3

Write down the formula: $d = \dfrac{m}{V}$

Substitute the values for m and V: $d = \dfrac{90}{9}$

Work out the answer and write down the units: $d = 10$ g/cm^3

So from the table on page 21, the statue is made of, what?

Of course an experiment of this type is never perfectly accurate, so the density that you measure will never be exactly the same as the official values.

REVIEW QUESTIONS

Q1 For each of the following objects, state whether they will sink or float or whether the outcome depends on the sample of material chosen:

a wood in oil
b wood in mercury
c plastic in oil
d steel in mercury
e silver in air

f gold in mercury (this experiment must be done rapidly as the gold will dissolve very quickly)
g helium balloon in air.

Q2 Write out the worked example on page 20 for the case of the student who measures all the lengths of the brick in m, and calculates with the mass in kg. Give the answer in kg/m^3.

Q3 A king believes that his jeweller has given him a crown that is a mixture of gold and silver, and not the 1.93 kg of pure gold that he paid for. He weighs the crown in a balance and finds that it has the correct mass of 1.93 kg. He then immerses it in a measuring jug where the water level was 800 cm^3. If the crown is pure gold, what will the new water level be? What will happen to the water level if the jeweller has cheated?

Examination questions are on page 50.

More questions on the CD ROM

5 FORCES

Videos & questions on the CD ROM

WHAT ARE FORCES?

A force is a push or a pull. The way that an object behaves depends on all of the forces acting on it. A force may come from the pull of a chain or rope, the push of a jet engine, the push of a pillar holding up a ceiling, and, as we have already seen, the pull of the gravitational field around the Earth.

Effects of forces

It is unusual for a single force to be acting on an object. Usually there will be two or more. The size and direction of these forces determine whether the object will move and the direction it will move in.

Forces are measured in **newtons**. They take many forms and have many effects including pushing, pulling, bending, stretching, squeezing and tearing. Forces can:
- **change the speed** of an object
- **change the direction** of movement of an object
- **change the shape** of an object.

To describe a force fully, you must state the size of the force and also the direction in which it is trying to move the object. The direction can be described in many different ways such as 'left to right', 'upwards' or 'north'. Sometimes it is useful to describe all of the forces in one direction as positive, and all of the forces in the other direction as negative. For two forces to be equal they must have the same size and the same direction.

WHAT IS FRICTION?

Friction is a very common force. It is the force that tries to stop movement between touching surfaces. Friction is caused by the roughness of the two surfaces, which produces resistance to movement.

In many situations friction can be a disadvantage, e.g. friction in the bearings of a bicycle wheel. In other situations, friction can be an advantage, e.g. between brake pads and a bicycle wheel.

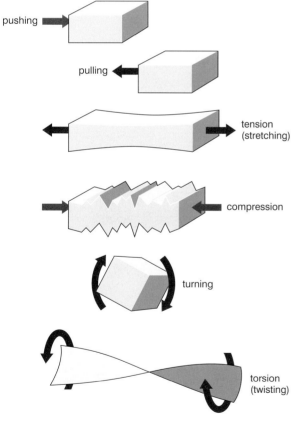

Combinations of forces can have all kinds of effects.

pushing

pulling

tension (stretching)

compression

turning

torsion (twisting)

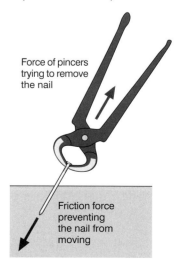

Force of pincers trying to remove the nail

Friction force preventing the nail from moving

Friction can stop any movement occurring at all, and it is friction that stops a nail coming out of a piece of wood.

ADDING FORCES

If two or more forces are pulling or pushing an object in the same direction, then the effect of the forces will add up; if they are pulling it in opposite directions, then the backwards forces can be subtracted.

Six husky dogs are pulling a sledge similar to the one in the picture. The sledge is travelling to the right and each dog is pulling with a force of 50 N. There is a friction force of 250 N that is trying to slow the sledge, and therefore must be pointing to the left.

The total force to the right is (6×50) N.
The total force to the left is 250 N.
The **resultant** force (the total added-up force) = 300 – 250 N to the right
= 50 N to the right.

Note that you must give the direction of the resultant force.

The sledge will behave as if this one force was acting on it, and therefore it will be accelerating to the right.

BALANCED FORCES

Usually there are at least two forces acting on an object. If these two forces are **balanced** then the object will either be stationary or moving at a constant speed.

A spacecraft in deep space will have no forces acting on it – no air resistance (no air), no force of gravity – and because there is no need to produce a forward force from its rockets, it will travel at a constant speed.

UNBALANCED FORCES

If the forces acting on an object are **unbalanced**, then it will change its speed or direction of movement – it will accelerate.

As a gymnast first steps on to a trampoline, his weight is much greater than the opposing supporting force of the trampoline, so he moves downwards, stretching the trampoline. As the trampoline stretches, its supporting force increases until the supporting force is equal to the gymnast's weight. When the two forces are balanced, the trampoline stops stretching. If an elephant stood on the trampoline, it would break because it could never produce a supporting force equal to the elephant's weight.

You see the same effect if you stand on snow or soft ground. If you stand on quicksand, then the supporting force will not equal your weight, and you will continue to sink.

The husky dogs are able to pull this sledge due to the low level of friction between the sledge and the snow.

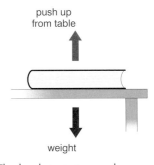

The book is stationary because the push upwards from the table is equal to the weight downwards. If the table stopped pushing upwards, the book would fall.

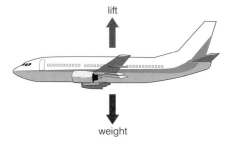

This aeroplane is flying 'straight and level' because the lift generated by the air flowing over the wings is equal and opposite to the weight of the aeroplane. This diagram shows that the plane will neither climb nor dive, as it would if the forces were not equal.

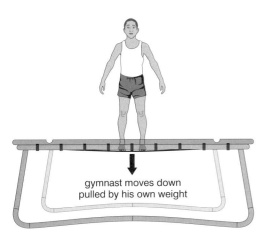

gymnast moves down pulled by his own weight

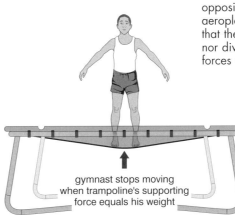

gymnast stops moving when trampoline's supporting force equals his weight

A trampoline stretches until it supports the weight on it.

The parachutist descends slowly, with the air resistance on the parachute causing it to pull upwards with a force exactly equalling the parachutist's weight.

As a skydiver jumps from a plane, the weight will be much greater than the opposing force caused by air resistance. Initially she will accelerate downwards at 10 m/s², just the same as the coconut on page 13.

The skydiver's speed will increase rapidly – and the force caused by the air resistance increases as the skydiver's speed increases. Eventually it will exactly match the weight, the forces will be balanced and the speed of the skydiver will remain constant. This speed is known as the **terminal speed**, typically 180 km/h.

If the skydiver makes herself streamlined by going headfirst, with her arms by her side, then she will cut through the air more easily, and the air resistance will be less. She will then accelerate again, until the force of air resistance increases again to equal her weight. She will now be going at almost 300 km/h.

A parachute has a very large surface area, and produces a very large resistive force, so the terminal speed of a parachutist is quite low. This means that he or she can land relatively safely.

HOW ARE MASS, FORCE AND ACCELERATION RELATED?

The acceleration of an object depends on its **mass** and the **force** that is applied to it. The relationship between these factors is given by the formula:

force = mass × acceleration	F = force in newtons
$F = m\,a$	m = mass in kg
	a = acceleration in m/s²

This equation explains the definition of the newton. 'A newton is the force that will accelerate a mass of 1 kg at 1 m/s².'

The equation is perhaps easier to understand if we rearrange it into the form $a = \frac{F}{m}$. This shows that if you use a big force you will get a larger acceleration, but if the object has more mass, then you get a smaller acceleration.

So a light object with a large force applied to it will have a large acceleration. (Think of an athlete with a racing bicycle.) But a massive object with a small force applied to it will have a small acceleration. (Think of a small child trying to pedal a large bicycle rickshaw.)

A* EXTRA

- The equation $F = m\,a$ shows that the acceleration of an object is directly proportional to the force acting on it (if its mass is constant) and is inversely proportional to its mass (if the force is constant).

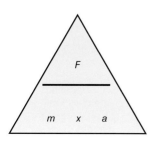

WORKED EXAMPLES

1 What force would be required to give a mass of 5 kg an acceleration of 10 m/s²?

Write down the formula:	$F = m\,a$
Substitute the values for m and a:	$F = 5 \times 10$
Work out the answer and write down the units:	$F = 50$ N

2 A car has a resultant driving force of 6000 N and a mass of 1200 kg.
   Calculate the car's initial acceleration.

Write down the formula in terms of a:	$a = \dfrac{F}{m}$
Substitute the values for F and m:	$a = \dfrac{6000}{1200}$
Work out the answer and write down the units:	$a = 5 \text{ m/s}^2$

MOTION IN A CIRCLE

If a moving object has no forces acting on it, it will continue to move in a
straight line at constant velocity.

So, if an object is moving in a circle, or along the arc of a circle, it follows
that there must be a force acting on it, to change its direction. Moving in
a circle means that the direction of motion is constantly changing, so this
in turn means that the direction of the force is constantly changing.

In order for the object to move on a circular path, the force must
always be acting towards the centre of the circle.

This force, which always acts towards the centre of the circle, is
given the name of **centripetal force**. You will also see that the
centripetal force also acts perpendicularly to the direction of
motion of the object at any instant.

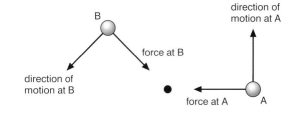

The centripetal force is not a new and different force from any you
have come across before, but is one or more of the forces already
acting on the object which is moving in the circle.

The table below gives some examples.

Example	How is the centripetal force supplied?
1. a stone on the end of a string, being whirled in a horizontal circle	by the tension force in the string
2. the Moon, orbiting the Earth	by the gravitational force of the Earth on the Moon
3. a car turning a corner	by the sideways friction force of the road on the tyres
4. a train going round a bend	by the sideways force of the rails on the wheels
5. a person standing on the Earth, which is spinning rapidly	by the gravitational force of the Earth on the person

See if you can think of other examples of things moving around arcs of
circles, and work out what force is providing the centripetal force.
Remember that the centripetal force is always towards the centre of the
arc and perpendicular to the direction in which the object is travelling at
that instant.

Consider a stone being whirled in a circle on the end of a string.

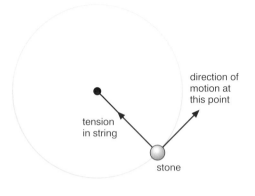

tension in string

direction of motion at this point

stone

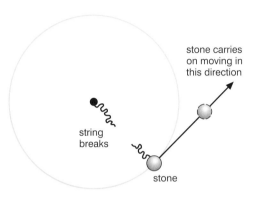

stone carries on moving in this direction

string breaks

stone

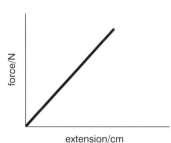

force/N

extension/cm

Elastic behaviour in a spring is shown by a straight line.

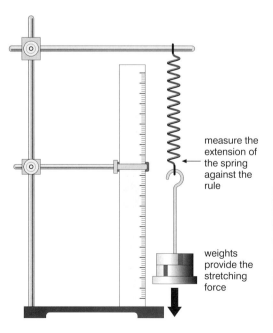

measure the extension of the spring against the rule

weights provide the stretching force

What happens if the string breaks?

In this case, the centripetal force is suddenly removed. There is now no force acting on the stone, so it continues to move in a straight line in whatever direction it had when the string broke (i.e. along the tangent to the circle at that point).

Of course, once it is free of the string, the effects of gravity start to change the motion, but initially the stone flies off along a tangent. It is important to note that the stone does NOT fly out along a radius. [A common mistake some people make is to say that there is a centrifugal (note: not centripetal) force pulling the stone outwards, which leads to the conclusion that when the string breaks, the stone flies out along a radius. This idea is incorrect.]

See if you can work out what would happen:
(a) to the Moon if gravity suddenly ceased,
(b) to a car turning a corner when the road was very slippery,
(c) to a person on Earth if gravity suddenly ceased.

HOOKE'S LAW

When a spring stretches, the extension of the spring is proportional to the force stretching it, provided the elastic limit of the spring is not exceeded. This is **Hooke's law** and is shown by a straight line on a graph.

The gradient of the line is a measure of the stiffness of the spring.

An experiment to measure Hooke's law
1 Assemble the apparatus shown on the left, but with only the mass hanger hooked on to the spring. The slotted masses are added later. For accurate readings, the mass hanger needs to be close to the rule, but not touching it.

2 Note and record the reading on the scale of the rule, next to the bottom of the mass hanger.

3 Add one slotted mass to the mass carrier. Slotted masses commonly used by schools are typically 100 g (weight 1 N) or 50 g (weight 0.5 N). Note and record the new scale reading of the bottom of the mass hanger.

4 Repeat step 3, adding masses one at a time and recording the corresponding scale readings. Add the masses carefully so that the spring stretches slowly.

5 You may want to reverse the experiment to see what happens as the masses are removed.

6 Calculate the extension of the spring caused each time by the load on the hanger.

The table below gives the results of one such experiment.

Mass/g	Force/N	Reading/cm	Calculate the extension/cm	Extension/cm
0	0	15.2	–	–
100	1.0	16.8	16.8 – 15.2	1.6
200	2.0	18.5	18.5 – 15.2	3.3
300	3.0	19.9	19.9 – 15.2	4.7
400	4.0	21.6	21.6 – 15.2	6.4
etc.	etc.			

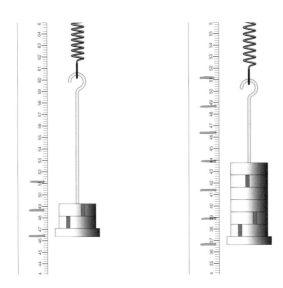

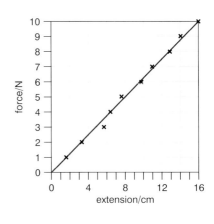

If any two quantities are proportional, when they are plotted against each other on a graph, the graph will have two characteristics:

(a) It will be a straight line,
(b) It will pass through the origin.

Because the graph of Force against Extension is a straight line which passes through the origin, this experiment verifies Hooke's law.

If you do stage 5 in the experiment above, you will also note that the spring, as well as showing proportional behaviour, also shows elastic behaviour – when the force is removed, the spring returns to its original length.

LIMIT OF PROPORTIONALITY

If you stretch the spring too far, the line is no longer straight, and Hooke's law is no longer true. This point at the end of the straight line is know as the 'limit of proportionality'.

The spring may (if you do not stretch it too far) be elastic and go back to its original length.

But as you stretch the material beyond the limit of proportionality, different materials can behave in widely different ways.

HOW ARE OTHER MATERIALS AFFECTED BY STRETCHING?

A music wire string, such as a guitar string, will behave as shown in the graph for wire (see page 30), but will break shortly after the limit of proportionality is reached.

A piece of rubber stretches quite a lot for small forces. The long polymer molecules are being 'straightened out'. Once this is done it becomes much stiffer and harder to extend further. However, unless it breaks, its behaviour is elastic.

A copper wire has a large plastic section on the graph. As it stretches, the wire becomes thinner and thinner until it finally breaks. This stretching is irreversible, and the extension is caused by 'plastic flow'.

A* EXTRA

- During elastic behaviour, the particles in the material are pulled apart a little. During plastic behaviour the particles slide past each other, and the structure of the material is changed permanently.

A strip of polythene will stretch relatively easily, but it will scarcely shorten at all when the load is removed. This means that polythene stretches almost entirely by plastic flow. This was the original meaning of the word 'plastic'. When people started to invent new materials in the early 1900s, many of them stretched in a plastic way. The word was then used to describe them.

Force–extension graphs for a metal wire, for rubber, for a polythene strip, and to show plastic flow.

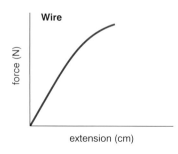

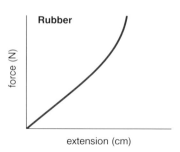

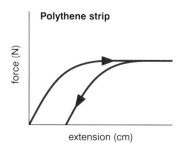

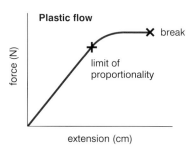

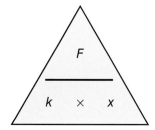

The equation for Hooke's law is:

force = stiffness of spring × extension of spring	F = force in newtons
$F = k \times x$	k = stiffness of the spring in N/m
	x = extension of the spring in m

Note that it is acceptable to use a stiffness in N/cm or N/mm, so long as the extension is measured in the same units.

This equation works for springs that are being stretched or compressed. The value of k will be the same for both, but note that some springs cannot be compressed (if, for example, the turns of the spring are already in contact).

WORKED EXAMPLE

A motorbike has a single compression spring on the rear wheels. When the cyclist sits on the bike, she pushes on the rear wheel with 60 per cent of her weight. If her mass is 50 kg, and the stiffness of the spring is 60 N/cm, how much does the spring compress when she sits on the bike?

Write down the formula for the weight of the cyclist: $W = m \times g$
Substitute the values for m and g: $W = 50 \times 10$
Work out the answer and write down the units: $W = 500$ N

The force on the rear spring = 60 per cent of 500 N
= 0.6 × 500 N
= 300 N

Write down the formula for the compression of the spring: $x = \dfrac{F}{k}$

Substitute the values for F and k: $x = \dfrac{300}{60}$

Work out the answer and write down the units: $x = 5$ cm

The spring compresses by 5 cm

Turning effect

If you have used a spanner to tighten a nut, or you have turned the handle of a rotary beater, you have used a force to turn something. But turning applies to less obvious examples, such as when you push the door handle to close a door, or when a child sits on the end of a see-saw to push her end of it down.

The turning effect of a force is called the **moment** of the force.

The moment of a force depends on two things:
• the size of the force
• the distance between the line of the force and the turning point, which is called the **pivot**.

We calculate the moment of force using this formula:

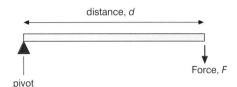

moment of a force = force × distance from pivot

moment = F d

Moment is measured in newton metres (Nm).

F = force in newtons (N)
d = distance in metres (m)

WORKED EXAMPLE

A girl pushes open a door with a force of 20 N. The door handle is at a distance of 0.80 m from the hinges.

Write down the formula:	moment = force × distance from pivot
Substitute the values for F and d:	moment = 20 N × 0.8 m
Work out the answer and write down the units:	moment = 16 Nm

WHEN MOMENTS BALANCE

The Principle of Moments says that if a system of forces is not turning, then the sum of the clockwise moments equals the sum of the anticlockwise moments about any point.

So, for example, if the following system of forces is balanced then:

sum of clockwise moments = sum of anticlockwise moments
moment of F_3 = moment of F_1 + moment of F_2
$(F_3 \times c) = (F_1 \times b) + (F_2 \times a)$

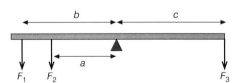

An experiment to verify the Principle of Moments
1 Drill a hole at the 50 cm mark of a metre rule.

2 Support the rule on a pivot through the drilled hole.

3 Using two loops of thread and two mass hangers and some slotted masses, suspend different weights, W_1 and W_2, at different distances, a and b, from the pivot. Carefully adjust the distances a and b until the rule balances horizontally.

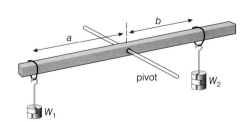

4 Record the values of W_1, W_2, a and b.

5 Repeat stages 3 and 4 several times, with different values of W_1, W_2, a and b.

6 For each set of results, calculate $(W_1 \times a)$ and $(W_2 \times b)$.

You will find that, within the limits of experimental accuracy, ($W_2 \times a$) and ($W_2 \times b$) will be equal for each set of readings.

	W_1/N	W_2/N	a/cm	b/cm	($W_1 \times a$)/Ncm	($W_2 \times b$)/Ncm
(a)	0.5	1.0	41.6	20.4	20.8	20.4
(b)	1.5	1.0	25.7	38.8	38.6	38.8
(c)	1.5	0.5	15.8	47.8	23.7	23.9
(d)	2.0	2.5	44.4	35.4	88.8	88.5

You will see that for each set of readings, the last two columns are equal, within the limits of the accuracy of the experiment. Thus the results verify the Principle of Moments.

The name we use in Physics to describe a set of balanced forces is **equilibrium**. When the system of forces is in equilibrium then the sum of the anticlockwise moments are balanced by the sum of the clockwise moments. In other words, there is no net moment on a body in equilibrium.

WORKED EXAMPLE

Two children are sitting on a see-saw. The see-saw is balanced on a pivot. Work out the boy's weight.

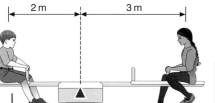

Forces on the see-saw

The girl is causing the clockwise moment of 400 N × 3 m.

The boy is causing the anticlockwise moment of W × 2 m.

The see-saw is balanced, so

the sum of the clockwise moments = the sum of the anticlockwise moments

$$400 \times 3 = W \times 2$$

$$W = 600 \text{ N}$$

WORKED EXAMPLE

A mechanic with a racing car team knows that a bolt on the engine is to be tightened to 60 Nm. If she is using a spanner that is 0.2 m long, with what force and in what direction should she push the end of the spanner?

Moment = force × distance from pivot

Rearrange the equation:

$$\text{force} = \frac{\text{moment}}{\text{distance from pivot}}$$

Substitute the values for moment and distance:

$$\text{force} = \frac{60 \text{ Nm}}{0.2 \text{ m}}$$

Work out the answer and write down the units: force = 300 N

This force should be applied at right angles to the end of the spanner.

Conditions for equilibrium

We use the word 'system' to describe a collection of objects working together. So in the example of the see-saw above, the two children and the see-saw form a system. We say that a system is in equilibrium if it is not moving in any direction and it is *not* rotating. We already know that for a system not to be moving, the forces on it must be equal and

So:

> For a system to be in equilibrium, there must be no resultant force and no resultant turning effect.

opposite.

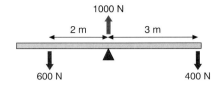
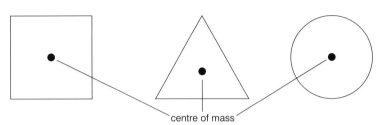

In the case of the balanced see-saw, we have already shown that there is no resultant turning effect on the see-saw because the clockwise and anticlockwise turning effects are equal and opposite. In addition, the downward weight of the two children on the see-saw is 1000 N, and the upward force on the see-saw from the pivot must also be 1000 N.

Centre of mass

The centre of mass is the point where we can assume *all* the mass of the object is concentrated. This is a useful simplification because we can pretend **gravity** only acts at a **single point** in the object, so a single arrow on a diagram can represent the **weight** of an object.

centre of mass

The centre of mass for objects with a regular shape is in the centre.

WHAT ABOUT IRREGULAR SHAPES?

To find the centre of mass of simple objects, such as a piece of card, follow these steps:

1 Hang up the object.
2 Suspend a mass from the same place.
3 Mark the position of the thread.
4 The centre of mass is somewhere along the line of the thread.
5 Repeat steps 1 to 3 with the object suspended from a different place.
6 The centre of mass is where the two lines cross.

card

retort stand

mass hanging on a thread

CENTRE OF MASS LINKS TO STABILITY

The idea of centre of mass is useful when predicting whether or not an

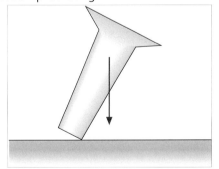

This object will topple over – a vertical line from the centre of mass falls outside the base of the object so the weight of the object tips it over the rest of the way. The moment of the force turns the object over. An object that is easy to topple is said to be in unstable equilibrium.

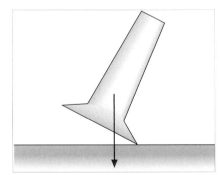

This object will fall back into place – a vertical line from the centre of mass falls inside the base of the object so the weight of the object pulls it back onto its base. The moment of the force returns the object to its base. An object that is difficult to topple is said to be in stable equilibrium.

Scalars and vectors

Force, velocity and acceleration are examples of **vector** quantities. A vector has a specific direction as well as a size, with a unit. We will meet many more later in this book: pressure, electrical current and even the flow of heat are all vectors.

Speed and mass are examples of a **scalar** quantity. A scalar quantity has size only, with a unit. There are many more scalar quantities to be met: temperature, work, power and electrical resistance are all scalars.

GRAPH TO FIND RESULTANT

If two husky dogs are pulling a sledge in different directions, it is clear that the sledge will move in a direction that is some sort of average of these directions. To find out exactly what will happen, we replace the two forces with a single force (the resultant) that will have just the same effect.

To calculate this single force we draw the two forces in the correct direction and to a scale length that is suitable. In the case of the husky dogs, a suitable scale might be 1 cm per 10 N, or perhaps 1 cm per 5 N.

We then find the resultant by completing a parallelogram. The resultant is then the diagonal line across the parallelogram between the two forces. This gives us the direction of the resultant force. We can calculate the magnitude of the resultant force by measuring its length, and using the scale that we chose to begin with.

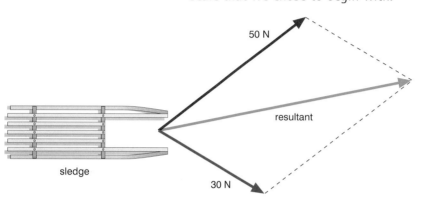

In this case one of the dogs is not working very hard, and the sledge will start to go in the direction in which the stronger dog is pulling.

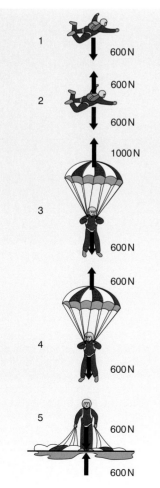

REVIEW QUESTIONS

Q1 The diagram shows the stages in the descent of a skydiver.
  a Describe and explain the motion of the skydiver in each stage.
  b In stage 5 explain why the parachutist does not sink into the ground.

Q2 The skydiver performed an experiment stretching a spring. She loaded masses onto the spring and measured its extension. Here are her results.

Extension/cm	0	4	8	12	16	20	24
Load/N	0	2.0	4.0	6.0	7.5	8.3	8.6

  a On graph paper, plot a graph of Load (*y*-axis) against Extension (*x*-axis).
    Draw a suitable line through your points.
  b Mark on the graph the limit of proportionality, and indicate the region where proportional behaviour occurs and the region where the behaviour is probably plastic.
  c How does the skydiver check whether the spring, after being loaded with 8.6 N has shown plastic behaviour or purely elastic behaviour?
  d Calculate the spring constant (stiffness) of the spring in the region of proportional behaviour. Note that you must give the units as well as the value.

Q3 The manufacturer of a car gave the following information:
Mass of car 1000 kg. The car will accelerate from 0 to 30 m/s
in 12 seconds.
   a Calculate the average acceleration of the car during the
     12 seconds.
   b Calculate the force needed to produce this acceleration.

Q4 Two tug boats have ropes attached to a ship and are about to
start moving it very carefully. One tug is north of the ship and is
pulling with a force of 3000 N, and the other tug is east of the
ship and is pulling with a force of 4000 N.
   a By means of a diagram calculate the total force with which the
     ship will be pulled, and show the direction in which it will be
     pulled.
   b If the ship has a mass of 500 tonnes (1 tonne = 1000 kg) how
     fast will it be moving 10 s after it starts?

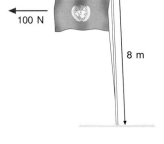

Q5 A flag is being blown by the wind. The force on the flag is
100 N and the flagpole is 8 m tall.
Calculate the moment of the force about the base of the flagpole.

Q6 Which of these glasses is the most stable? Explain your answer.

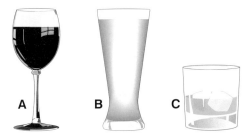

Examination questions are on page 50.

More questions
on the CD ROM

6 ENERGY, WORK AND POWER

Energy

If you own a car, it will not move without fuel. At present this fuel could be petrol, or alcohol or diesel fuel. In the past the fuel could, just possibly, have been coal; and in the future it could be hydrogen, or electricity stored in a battery. But whatever fuel you use, you are buying something with the ability to make that car move. This stored ability is known as potential energy.

A clock needs energy to make the hands move, and this energy can be stored in a spring that you wind up with a key, in an electrical battery, or in weights that are raised up.

Fuel, whatever form it comes in, gives a car the ability to move.

These electric streetcars take electrical energy from the overhead wire and convert it into kinetic energy.

Stored, or hidden, energy is called **potential energy** (p.e.). In this context 'potential' does not mean 'with qualities that may lead to future success' (as in 'potential film star') but rather 'containing power'. If a spring is stretched, the spring will have potential energy. If a load is raised above the ground, it will have **gravitational potential energy**.

If the spring is released or the load moves back to the ground, the stored potential energy is transferred to movement energy, which is called **kinetic energy** (k.e.).

In all of the examples above, the potential energy can be used to make an object move, and hence give it kinetic energy. Kinetic energy can also be turned into potential energy, and this can be seen most clearly in the action of a pendulum, where at each end of its swing the pendulum has a maximum amount of gravitational potential energy, and at the middle of its swing where some of the potential energy has been used (the pendulum is lower down) and turned into kinetic energy (the pendulum is moving fastest).

Gravitational potential energy can be calculated using the formula:

gravitational potential energy = mass × gravitational field strength × height

$$p.e. = m\,g\,h$$

p.e. = gravitational potential energy in joules (J)

m = mass in kilograms (kg)

g = gravitational field strength of 10 N/kg

h = height in metres (m)

WORKED EXAMPLE

A skier has a mass of 70 kg and travels up in a ski lift a vertical height of 300 m. Calculate the change in the skier's gravitational potential energy.

Write down the formula:	$p.e. = m\,g\,h$
Substitute values for m, g and h:	$p.e. = 70 \times 10 \times 300$
Work out the answer and write down the unit:	$p.e. = 210\,000$ J or 210 kJ

KINETIC ENERGY

The kinetic energy of an object depends on its mass and its speed. The kinetic energy can be calculated using the following formula:

kinetic energy = $\frac{1}{2}$ × mass × speed^2

$$k.e. = \tfrac{1}{2}mv^2$$

k.e. = kinetic energy in joules (J)

m = mass in kilograms (kg)

v = speed in m/s

WORKED EXAMPLE

An ice skater has a mass of 50 kg and travels at a speed of 5 m/s. Calculate the ice-skater's kinetic energy.

Write down the formula:	$k.e. = \tfrac{1}{2}mv^2$
Substitute the values for m and v:	$k.e. = \tfrac{1}{2} \times 50 \times 5 \times 5$
Work out the answer and write down the unit:	$k.e. = 625$ J

POTENTIAL ENERGY

As can be seen in the pendulum, energy can either be stored or can be seen in motion in some way.

The different types of stored energy are all forms of potential energy. Here are some important examples:

A* EXTRA

- As a skier skis down a mountain the loss in potential energy should equal the gain in kinetic energy (assuming no other energy transfers take place, as a result of friction, for example). Calculations can then be performed using:

  Loss in p.e. = Gain in k.e.
  $(mgh = \tfrac{1}{2}mv^2)$

Gravitational energy or gravitational potential energy
This is energy stored by an object being raised up in a gravitational field, for example, a ball on top of a hill.

Strain energy
The word 'strain' means stretched. Strain energy can be stored in springs (in clocks, for example) and in bows when they are drawn back before the arrow is released.

Chemical energy
In any object the atoms are held together by forces that are called bonds. These bonds behave like springs. In some materials, such as fuels and explosives, the bonds are forced to be shorter or longer than they wish. This stores energy in the bonds that can be released by breaking up the structure of the fuel or the explosive.

A battery is ready to turn chemical energy into electrical energy, and a rechargeable battery is so called because every time that it is discharged it can be recharged by forcing electricity through it backwards. The electrical energy that is fed is stored as chemical energy.

Nuclear energy
The energy in a nucleus of an atom is stored in the extremely strong bonds between the particles of which the nucleus is made. Some of this energy can be released, in the case of uranium (and a couple of other metals) by splitting the nucleus of the atom into two smaller nuclei. This can be done either slowly and for good purposes in a nuclear power station, or very rapidly in an atomic bomb.

Here are some other important types of energy. They are actually all different sorts of kinetic energy, but this is far from obvious in some cases:

If people just use the words 'kinetic energy', then they are referring to the energy of a visible moving object with k.e. = $\frac{1}{2}mv^2$.

Internal energy
This is contained within an object and makes the difference between the object being hot or cold. As we see on page 56, a hot object contains atoms that are moving fast or vibrating strongly.

Electrical energy
Electrical currents carry electrical energy from one place to another. Electrical energy can easily be turned into kinetic energy in a motor or internal energy in a resistor, perhaps used as a heater.

Light energy
Light carries light energy as it travels. This will be turned into internal energy if it strikes most objects, but it can be made to generate electrical energy if it hits a solar panel.

Sound waves
These carry a very small amount of energy from the source of the noise. (Do not confuse the 2000 W of electricity consumed by the equipment of a rock group performing on stage, with the 100 W of sound being emitted by the loudspeakers. The ear is extremely good at detecting sound.)

CONVERSION OF ENERGY

We have already explained how kinetic energy and gravitational potential energy can be converted backwards and forwards. In fact any type of energy can be converted into any other type of energy. In some cases this conversion can be done efficiently, such as between kinetic energy and electrical energy. In other cases the conversion is inefficient, the most notorious example of this being the power station where only between 40 and 60 per cent of the internal energy in the fuel is converted to electricity.

In every case of conversion of energy, some of the energy is converted to internal energy. The light bulb creating light energy from electrical energy gets hot; the electric motor turning electrical energy to kinetic energy gets hot; the diesel engine using chemical energy gets hot, the battery that is being charged gets hot. Even the pendulum eventually stops swinging because the movement of the pendulum through the air heats up the air.

CONSERVATION OF ENERGY

Don't confuse the words **conversion** and **conservation**.

Energy cannot be created or destroyed. You may need to describe how energy is transferred in different situations, but remember that total energy is always conserved: the total energy at the start and at the end must have the same total value.

So you must account for all of the energy, and that includes the internal energy that will have been created, as well perhaps as light or sound.

For example, the tram takes electrical energy and converts it mainly into kinetic energy, but also into internal energy and sound.

You can use the principle of the conservation of energy to calculate what happens when kinetic energy and potential energy are converted from either one to the other.

So long as negligible energy is lost in the conversion, $mgh = \frac{1}{2}mv^2$.

WORKED EXAMPLE

If a stone thrown vertically upwards reaches a height of 6 m above the hand of the thrower, with what speed was it thrown?

The decrease in k.e. of the stone as it rises = the increase in the p.e. of the stone.

As the final k.e. of the stone = 0, the initial k.e. of the stone = the increase in the p.e. of the stone at the top of its flight.

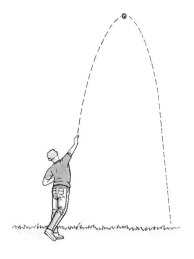

The kinetic energy given to the stone when it is thrown is transferred to potential energy as it gains height and slows down. At the top of its flight practically all the kinetic energy will have been converted into gravitational potential energy. A small amount of energy will have been lost due to friction between the stone and the air.

Write down the formula:	$\frac{1}{2}mv^2 = mgh$
Note that the mass has cancelled out; the mass does not matter in this case.	$\frac{1}{2}v^2 = gh$
Substitute values for g and h:	$v^2 = gh \times 2$
	$= 10 \times 6 \times 2$
	$= 120$
Work out the answer and write down the unit:	$v = \sqrt{120}$
	$= 10.95$ m/s

Energy resources

Most of the energy we use is obtained from **fossil fuels** – coal, oil and natural gas.

Once supplies of these fuels have been used up they cannot be replaced – they are **non-renewable**.

At current levels of use, oil and gas supplies will last for about another 40 years, and coal supplies for about a further 300 years. The development of **renewable** sources of energy is therefore becoming increasingly important.

The **wind** is used to turn windmill-like turbines which generate electricity directly from the rotating motion of their blades. Modern wind turbines are very efficient but you would need several thousand to equal the generating capacity of a modern fossil-fuel power station.

The motion of **waves** can be used to move large floats and generate electricity. A very large number of floats is needed to produce a significant amount of electricity.

Dams on tidal estuaries trap the water at high tide. When the water is allowed to flow back at low tide, it can turn turbines to drive electrical generators. This obviously limits the use of the estuary.

On a windy day this wind turbine generates 2000 kW of electricity. That's enough for 1200 families.

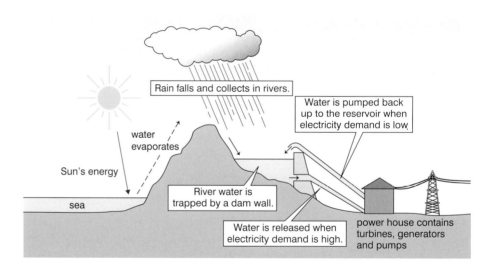

Rain falls and collects in rivers.

Water is pumped back up to the reservoir when electricity demand is low.

water evaporates

Sun's energy

sea

River water is trapped by a dam wall.

Water is released when electricity demand is high.

power house contains turbines, generators and pumps

A 'pumped storage' hydroelectric power station.

Dams can be used to store **water** which is allowed to fall in a controlled way that generates electricity. This is particularly useful in mountainous regions for generating **hydroelectric power**. When demand for electricity is low, spare electricity from a nuclear power station (which has to be run continuously) can be used to pump water back up into the high reservoir for use in times of high demand.

Plants use energy from the Sun in photosynthesis. Plant material can then be used as a **biomass fuel** – either directly by burning it or indirectly. A good example of indirect use is to ferment sugar cane to make ethanol, which is then used as an alternative to petrol. Waste plant material can be used in 'biodigesters' to produce methane gas. The methane is then used as a fuel.

Geothermal power is obtained using the heat of the Earth. In certain parts of the world, water forms hot springs which can be used directly for heating. Water can also be pumped deep into the ground to be heated.

SUN – NUCLEAR FUSION

Solar power is energy from the Sun, which itself is powered by nuclear fusion reactions (where the small nuclei of hydrogen atoms join to make larger nuclei that are, in fact, helium and an enormous amount of energy is released.

The Sun's energy is trapped by solar panels and transferred into electrical energy or, as with domestic solar panels, is used to heat water. The cost of installing solar panels is high, but in sunny countries solar power is of increasing importance.

Work

Work is done when the application of a force results in movement. Work can only be done if the object or system has energy. When work is done energy is transferred.

Work done can be calculated using the following formula:

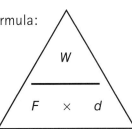

work done = force × distance moved

$W = F\,d = \Delta E$

W = work done in joules (J)

F = force in newtons (N)

d = distance moved in the direction of the force in metres (m)

ΔE = energy, transferred in joules (J)

In this position the gymnast is not doing any work against his body weight – he is not moving (he will be doing work pumping blood around his body though).

DELTA NOTATION

We use the Greek letter Δ (delta) to stand for 'the change in'. For example, ΔE means 'the change in the energy'. When you are using ΔE in an equation, treat them as one symbol meaning 'the change in energy'; so don't even think of separating them.

WORKED EXAMPLES

1 A cyclist pedals along a flat road. She exerts a force of 60 N and travels 150 m. Calculate the work done by the cyclist.

Write down the formula:	$\Delta W = F\,d$
Substitute the values for F and d:	$\Delta W = 60 \times 150$
Work out the answer and write down the unit:	$\Delta W = 9000$ J

2 A person does 3000 J of work in pushing a supermarket trolley 50 m across a level car park. What force was the person exerting on the trolley?

Write down the formula with F as the subject:	$F = \dfrac{\Delta W}{d}$
Substitute the values for ΔW and d:	$F = \dfrac{3000}{50}$
Work out the answer and write down the unit:	$F = 60$ N

The gymnast is doing work. He is moving upwards against his weight. Energy is being transferred as he does the work.

Power

A powerful engine in a car can take you up a road to the top of a mountain more quickly than a less-powerful engine. Both engines can do the work, given enough time, but the powerful engine can do the work more quickly. In the same way, a powerful electric motor on a cooling fan will move the air in the room more quickly; and the 'powerfully built' athlete will, by transferring more kinetic energy to it as it is launched, throw the javelin further.

Power is defined as the rate of doing work or the rate of transferring energy. The more powerful a machine is, the quicker it does a fixed amount of work or transfers a fixed amount of energy.

Power can be calculated using the formula:

$$\text{power} = \frac{\text{work done}}{\text{time taken}} = \frac{\text{energy transfer}}{\text{time taken}}$$

$$P = \frac{W}{t} \text{ or } P = \frac{E}{t}$$

P = power in joules per second or watts (W)

E = energy transferred in joules (J)

W = work done in joules (J)

t = time taken in seconds (s)

WORKED EXAMPLES

1 A crane lifts a 100 kg girder for a skyscraper by 20 m in 40 s. Hence it does 20 000 J of work in 40 seconds. Calculate its power over this time. Note: this calculation tells you the size of electric motor that the crane needs.

Write down the formula:	$P = \dfrac{W}{t}$
Substitute the values for W and t:	$P = \dfrac{20\,000}{40}$
Work out the answer and write down the unit:	$P = 500\,\text{W}$

2 A student with a weight of 600 N runs up the flight of stairs shown in the diagram (left) in 4 seconds. Calculate the student's power.

Write down the formula for work done:	$W = F\,d$
Substitute the values for F and d:	$W = 600 \times 5 = 3000\,\text{J}$
Write down the formula for power:	$P = \dfrac{W}{t}$
Substitute the values for W and t:	$P = \dfrac{3000}{4}$
Work out the answer and write down the unit	$= 750\,\text{W}$

The student is lifting his body against the force of gravity, which acts in a vertical direction. The distance measured must be in the direction of the force (that is, the vertical height).

REVIEW QUESTIONS

Q1 **a** What is meant by a non-renewable energy source?
  b Name three non-renewable energy sources.
  c Which non-renewable energy source is likely to last the longest?

Q2 Look at the graphs, which show the amount of energy from different sources used in the OECD nations between 1980 and 2000 and the world trends of energy consumption.

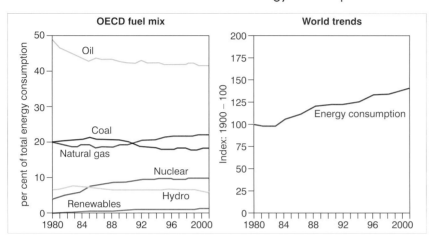

  a Describe the trend in the total energy used in this period.
  b Describe the main changes in the sources of this energy.
  c What do you expect to happen to these graphs in the next 20 years? Give reasons where possible.

Q3 A site has been chosen for a wind farm (a series of windmill-like turbines).
  a Give two important factors in choosing the site.
  b Give one advantage and one disadvantage of using wind farms to generate electricity.

Q4 50 000 J of work are done as a crane lifts a load of 400 kg. How far did the crane lift the load? (Gravitational field strength, g, is 10 N/kg.)

Q5 A student is carrying out a personal fitness test.
She steps on and off the 'step' 200 times.
She transfers 30 J of energy each time she steps up.
  a Calculate the energy transferred during the test.
  b She takes 3 minutes to do the test. Calculate her average power.

Q6 A child of mass 35 kg climbed a 30 m high snow-covered hill.
  a Calculate the change in the child's potential gravitational energy.
  b The child then climbed onto a lightweight sledge and slid down the hill. Calculate the child's maximum speed at the bottom of the hill. (Ignore the mass of the sledge.)
  c Explain why the actual speed at the bottom of the hill is likely to be less than the value calculated in part b.

More questions on the CD ROM

Examination questions are on page 50.

7 PRESSURE

Videos & questions on the CD ROM

The snowmobile in the picture can travel over soft snow because its weight is spread by the skis over a large area of snow. If the rider got off and stood on the snow, he would probably sink into it up to his knees, even though he is much lighter than the snowmobile.

If a pair of shoes has small heels, the wearer can easily damage a wooden floor by sinking into it. And a drawing pin is pushed into a notice board by the pressure of your thumb.

In every case the question is not just what force is used, but also what area it is spread over. Where there is a large force over a small area, we have a high pressure, and a small force over a large area gives us a low pressure.

Pressure is measured in newtons per square metre (N/m^2), but note that it is often given the special name of pascal (Pa).

In order to measure how 'spread out' a force is, use this formula:

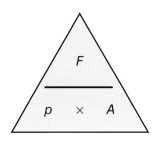

$$p = \frac{F}{A}$$

p = pressure in pascals, Pa (or newtons per square metre, N/m^2)

F = force in newtons, N

A = area in m^2

Note: $1 \text{ Pa} = 1 \text{ N/m}^2$

WORKED EXAMPLE

What pressure on the snow does the snowmobile make if it has a weight of 800 N and the runners have an area of 0.2 m²?

Write down the formula:	$p = \dfrac{F}{A}$
Confirm that F is in N and A is in m²	
Substitute the values for F and A:	$p = \dfrac{800}{0.2}$
Work out the answer and write down the units:	$p = 4000$ Pa
	$= 4$ kPa

Note that 4 kPa is a very low pressure. If you stand on the ground in basketball shoes, the pressure on the ground will be around 20 kPa. The wheel of a car generates a pressure on the ground of around 200 kPa. Pressures can be quite high, and so the kPa is often used.

ATMOSPHERIC PRESSURE

Because we have spent all of our lives living in the atmosphere of the Earth, we seldom think that we have 20 km or so of air pressing on us. We do not feel the pressure because it does not just push down, it pushes us inwards from all sides. Our lungs do not collapse, because the same air pressure flows into our lungs and presses outwards. It would be a very different story if our lungs did not contain any air and there was a vacuum inside them.

We can show this by seeing how a plastic bottle collapses if the air is removed from it.

The plastic bottle is filled with steam from a kettle so that the air in it is replaced by the steam. The lid is screwed onto the bottle, and the bottle is cooled by immersing it in cold water. When the steam turns back into water, the bottle collapses due to the pressure of the air outside.

Warning: steam from a kettle is extremely dangerous, and much more so than boiling water. You must never put your hands near the steam coming out of a boiling kettle, and you must not perform this experiment without suitable safety equipment and training.

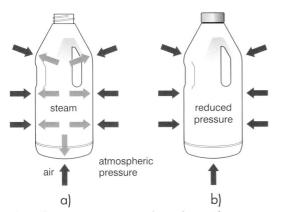

a) b) c)

a) Bottle with same pressure inside and out. The pressure inside is from steam at atmospheric pressure.
b) Bottle with internal pressure removed. You can do this with a strong metal bottle, but not with a plastic one.
c) Plastic bottle collapsed.

Atmospheric pressure is approximately 100 kPa. This value is a pure coincidence. In fact it is around 101.3 kPa, though it increases and decreases by 5 per cent or so depending on the weather. But in the same way that we often take g to be 10 m/s^2 on the Earth when it is more accurately 9.8 m/s^2, we often choose to take atmospheric pressure to be 100 kPa.

Pressure is also measured in bar and millibar. Normal atmospheric pressure is approximately 1 bar. The pressure on a scuba diver's cylinder of air can easily be 200 bar. You will see millibar used in some weather forecasts. Atmospheric pressure is approximately 1000 mbar.

THE MERCURY BAROMETER

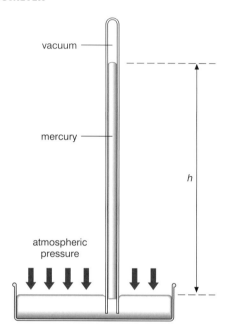

The mercury barometer is made of a glass tube, sealed at the top. It contains mercury, and the base of the tube dips into a beaker, and below the surface of the mercury in the beaker.

Atmospheric pressure pushes down on the mercury in the beaker, which in turn pushes mercury up the tube.

If the space above the mercury in the tube is a vacuum, then nothing is pushing down on the top of the mercury in the tube, and atmospheric pressure will push the mercury up until the pressure of the column of mercury balances the atmospheric pressure. The height h from the top of the mercury in the beaker, to the top of the mercury in the tube can be used to calculate atmospheric pressure.

This height h is approximately 760 mm of mercury, and in some countries atmospheric pressure is still quoted in mm of mercury.

Note that mercury has a convex (curved-upwards) shape when in contact with glass. You should measure to the top of the mercury in the tube, and to the flat surface of the mercury in the beaker.

Mercury barometers are no longer made because mercury is a highly poisonous metal with a poisonous vapour.

PRESSURE AND DEPTH

If you dive below the water, the height of the water above you also puts pressure on you. At a depth of 10 m of water, the pressure has increased by 100 kPa, and for each further 10 m of depth the pressure increases by another 100 kPa. The rapid increase in pressure explains why scuba divers cannot go down more than 20 m without great difficulty.

The hull of a submarine is made very strong so that the submariners can breathe air at the normal pressure.

The increase in pressure below the surface of a liquid depends on (a) the depth below the surface and (b) the density of the liquid. So the pressure will be much higher at a certain depth below the surface of mercury than it is below the surface of water. It does not depend on anything else, and note in particular that the pressure does not depend on the width of the water. If a diver goes to inspect a well, the pressure 10 m below the surface is the same as the pressure 10 m below the surface of a large lake. This explains why an engineer who is designing a dam needs to make it the same thickness whether the lake that has to be held back is 100 km long or only 100 m long.

These scuba divers breathe compressed air at high pressure to prevent their lungs collapsing due to the high pressure from the water above them. This is a safe sport, but only because novices are trained to a very high standard.

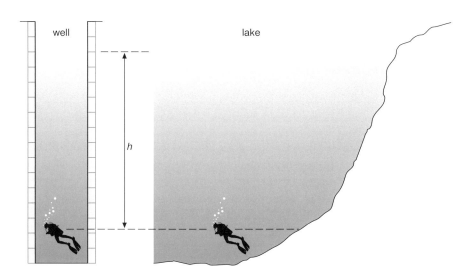

The pressure on the diver is the same in the well and in the lake. In both cases it depends only on the density of the liquid and his depth, *h*.

The pressure below the surface of a fluid, and in fact between any two points in the fluid can be calculated by the following equation:

pressure difference = height × density × strength of the gravitational field

$$p = h \times d \times g$$

p = pressure difference in pascals (Pa)

d = density in kilograms per cubic metre (kg/m^3)

g = acceleration of free fall (m/s^2)

Note that the density d must be in kg/m^3. If it is quoted in g/cm^3, you must convert it.

Scientists often use the Greek letter ρ for density, and write the above equation as

$$p = h\rho g$$

Note that there is one major cause of confusion. Consider the pressure on a scuba diver. Before he jumps in, the pressure on him is already 100 kPa (or 1 bar). When he has dived down 10 m, the pressure on him increases by 100 kPa, so the total pressure on him is now 200 kPa (2 bar). The pressure is coming 100 kPa from the air above him, and 100 kPa from the water above him. At 20 m, the total pressure on him is 300 kPa, and so on.

WORKED EXAMPLE

An aquarium in a 'sea life' centre has a tunnel through a tank of water at a depth of 5 m below the surface. The manufacturer guarantees the tunnel to a pressure difference of 200 kPa. Is the tunnel safe?

Write down the equation:	$p = h \times d \times g$
Substitute the values into the equation:	$p = 5 \times 1000 \times 10$
Work out the answer and write down the unit:	$p = 50\ 000$ Pa
	$= 50$ kPa.
The tunnel is safe.	

Note that the total pressure on the outside of the tunnel is 50 kPa from the water, plus 100 kPa from the air pushing on top of the water, giving 150 kPa. However, the tunnel is also full of air, which is pushing outwards with a pressure of 100 kPa. So the tunnel only has to stand a pressure difference of 50 kPa.

THE MANOMETER

Manometers are used to measure the pressure difference between two regions. They are used, for example, on cleanrooms where computer chips and other semiconductor devices are made. It is necessary to ensure that the pressure inside the room is slightly higher than outside in order to prevent dust finding its way into the cleanroom through small gaps in the wall.

A manometer is mounted on the wall of the cleanroom. It consists of a tube of plastic or glass, bent into the U shape shown, and filled with a liquid that is often oil. If there is a pressure difference between the ends of the manometer, the liquid moves until the pressure difference is balanced by the *difference* in height of the ends of the liquid. The greater the pressure, the greater the difference in height. You will note immediately that the liquid will be blown out if the pressure difference is too great.

Oil is often used rather than water because water evaporates and also because oil, being less dense, makes the manometer more sensitive: for the same pressure difference, the oil will move further.

The pressure difference between the two regions is given by the following equation:

$\Delta p = hdg$

This is basically the same equation as $p = h \times d \times g$.

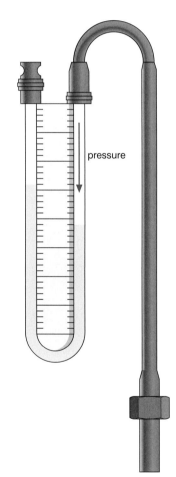

pressure

This manometer is used to measure the pressure in the supply of heating gas to a building. The U tube is filled with water and the height difference is measured. The pressure should be about 3000 Pa above atmospheric pressure, which is a height difference of 0.3 m. The gas won't burn properly if the pressure is wrong.

WORKED EXAMPLE

The manometer on an industrial machine shows that the oil is being pushed towards the machine and the height difference, h is 20 cm. What is the pressure inside the machine if the pressure outside is 100 kPa? The oil has a density of 800 kg/m^3.

Write down the equation:	$\Delta p = hdg$
Substitute the values into the equation:	$\Delta p = 0.2 \times 800 \times 10$
Be sure to convert the height, h into m:	20 cm = 0.2 m
Work out the answer and write down the unit:	$\Delta p = 1600$ Pa
	$= 1.6$ kPa

The pressure inside the machine must be lower than atmospheric pressure. Therefore the pressure inside the machine = $(100 - 1.6)$ kPa

$$= 98.4 \text{ kPa}$$

A manometer.

REVIEW QUESTIONS

Q1 Calculate the pressure generated by an ordinary shoe heel (person of mass 40 kg, heel 5 cm x 5 cm), an elephant (of mass 500 kg, foot of 20 cm diameter) and a high-heeled shoe (person of mass 40 kg, heel of area 0.5 cm^2). Which ones will damage a wooden floor that starts to yield at a pressure of 4000 kPa?

Note that to convert from cm^2 to m^2 you need to divide by 10 000.

Q2 The pressure gauge on a submarine in a river was reading 100 kPa when it was at the surface. If a sailor notices that the gauge is now reading 250 kPa, how deep is he? How would the answer change if he were diving in seawater that is slightly denser than fresh water?

More questions on the CD ROM

Examination questions are on page 50.

EXAMINATION QUESTIONS

Q1 Fig. 1.1 shows a cycle track.

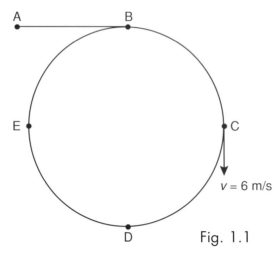

Fig. 1.1

A cyclist starts at A and follows the path ABCDEB.
The speed-time graph is shown in Fig. 1.2.

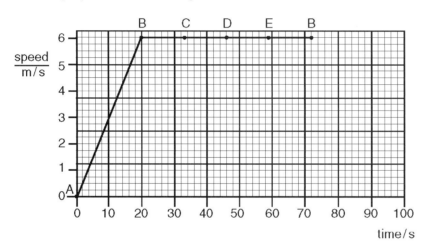

Fig. 1.2

a Use information from Fig. 1.1 and Fig. 1.2 to describe the motion of the cyclist
   i along AB,

   _____

   ii along BCDEB.

   _____

   _____ [4]

b The velocity v of the cyclist at C is shown in Fig. 1.1.
   State one similarity and one difference between the velocity at C and the velocity at E.
   similarity _____

   difference _____ [2]

c Calculate
   i the distance along the cycle track from A to B,
      distance = _____
   ii the circumference of the circular part of the track.
      circumference = _____ [4]

Q2 Fig. 2.1 shows the velocity-time graph for a bus during tests.
At time $t = 0$, the driver starts to brake.

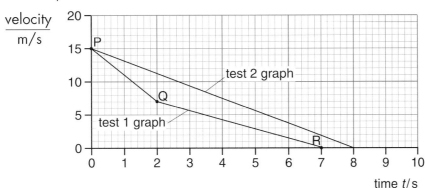

Fig. 2.1

a For test 1,
   i determine how long the bus takes to stop,

   _____

   ii state which part of the graph shows the greatest deceleration,

   _____

   iii use the graph to determine how far the bus travels in the first 2 seconds.
       distance = _____ [4]

b For test 2, a device was fitted to the bus. The device changed the deceleration.
   i State two ways in which the deceleration during test 2 is different from that during test 1.
       1 _____
       2 _____
   ii Calculate the value of the deceleration in test 2.
       deceleration = _____ [4]

c Fig. 2.2 shows a sketch graph of the magnitude of the acceleration for the bus when it is travelling around a circular track at constant speed.

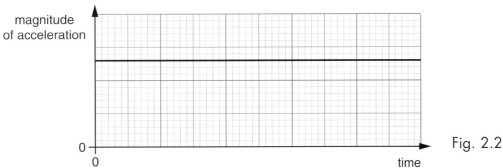

Fig. 2.2

   i Use the graph to show that there is a force of constant magnitude acting on the bus.

   _____

   _____

   ii State the direction of this force.

   _____ [3]

Q3 A large spring is repeatedly stretched by an athlete to increase the strength of his arms. Fig. 3.1 is a table showing the force required to stretch the spring.

extension of spring/m	0.096	0.192	0.288	0.384
force exerted to produce extension/N	250	500	750	1000

Fig. 3.1

a i State Hooke's law.

_____ [1]

ii Use the results in Fig. 3.1 to show that the spring obeys Hooke's law.

_____

_____ [1]

b Another athlete using a different spring exerts an **average** force of 400 N to enable her to extend the spring by 0.210 m.
  i Calculate the work done by this athlete in extending the spring once.
    work done = _____
  ii She is able to extend the spring by this amount and to release it 24 times in 60 s. Calculate the power used by this athlete while doing this exercise.
    power = _____ [4]

Q4 A group of students attempts to find out how much power each student can generate. The students work in pairs in order to find the time taken for each student to run up a flight of stairs. The stairs used are shown in Fig. 4.1.

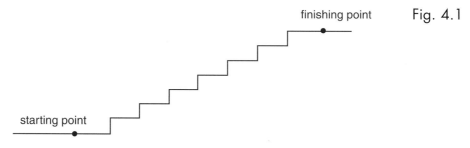

finishing point   Fig. 4.1

starting point

a Make a list of all the readings that would be needed. Where possible, indicate how the accuracy of the readings could be improved.

_____

_____

_____

_____

_____

_____ [4]

b Using words, not symbols, write down all equations that would be needed to work out the power of a student.

_____

_____

_____ [2]

c i When the student has reached the finishing point and is standing at the top of the stairs, what form of energy has increased to its maximum?

_____

ii Suggest why the total power of the student is greater than the power calculated by this method.

_____ [3]

Q5 Fig. 5.1 shows a diver 50 m below the surface of the water.

a The density of water is 1000 kg/m^3 and the acceleration of free fall is 10 m/s^2.
Calculate the pressure that the water exerts on the diver.
pressure = _____ [3]

b The window in the diver's helmet is 150 mm wide and 70 mm from top to bottom.
Calculate the force that the water exerts on this window.
force = _____ [3]

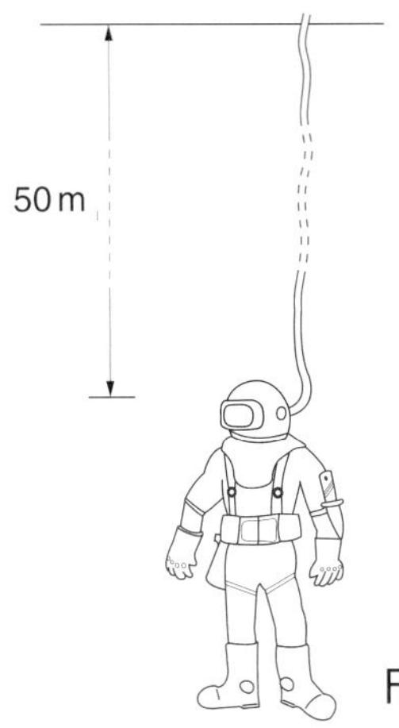

Fig. 5.1

Q6 Fig. 6.1 shows a smooth metal block about to slide down BD, along DE and up EF. BD and DE are friction-free surfaces, but EF is rough. The block stops at F.

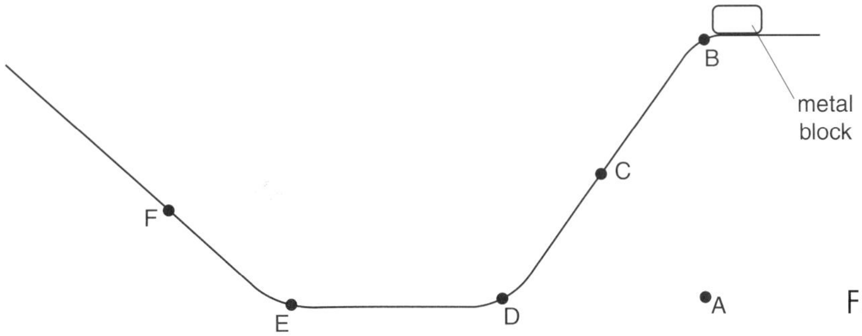

Fig. 6.1

a On Fig. 6.2, sketch the speed-time graph for the journey from B to F. Label D, E and F on your graph. [3]

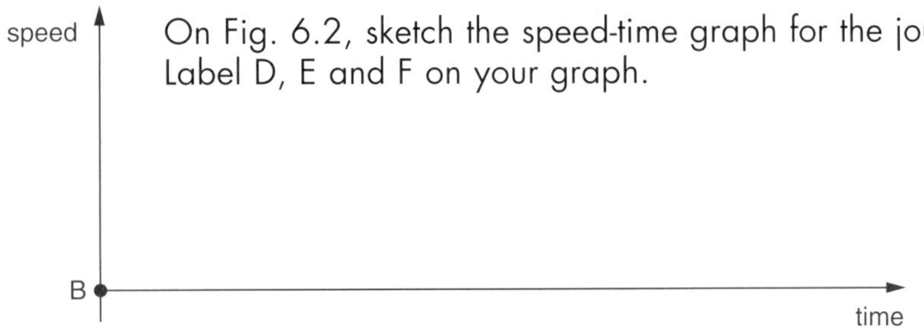

Fig. 6.2

b The mass of the block is 0.2 kg. The vertical height of B above A is 0.6 m. The acceleration due to gravity is 10 m/s^2.
   i Calculate the work done in lifting the block from A to B.
   work done = _____
   ii At C, the block is moving at a speed of 2.5 m/s. Calculate its kinetic energy at C.
   kinetic energy _____ [5]

c As it passes D, the speed of the block remains almost constant but the velocity changes. Using the terms *vector* and *scalar*, explain this statement.

_____

_____

_____ [2]

d F is the point where the kinetic energy of the block is zero. In terms of energy changes, explain why F is lower than B.

_____

_____

_____

_____ [3]

Many plants, such as the cactus, have very few
leaves, which reduces the surface area and so
fewer particles of water can evaporate. Desert
plants often have very long roots. These allow the
plant to store the water it has as deep as
possible, thus reducing the chances of the water
particles being lost as water vapour

Collection of succulents including:
Blue Echeveria, Panda Plant,
Haworthia Cymbiformis

Surviving the desert

What makes a desert? A place that is very hot? No, the key feature that defines a desert is a place that is
very dry. This might be because the area is close to the equator and so receives the most heating from the
Sun, or it could be because local mountain ranges cause rain to fall elsewhere, leaving a dry desert area.

If a plant or animal is going to survive in a desert area, particularly a hot desert area, it is going to
need to be careful about keeping and using water. High temperatures will cause any available water
to evaporate as some particles will have enough energy to leave the liquid. The lack of water means that
temperatures often fall very quickly at night. This is because invisible water vapour in the air usually
reflects the heat back down and so acts as a blanket to keep the ground warm, but above a desert there is
less water, and so it gets cold. Clouds also act as blankets, and deserts tend not to have these either.

THERMAL PHYSICS

Molecular graphic of evaporating water molcules

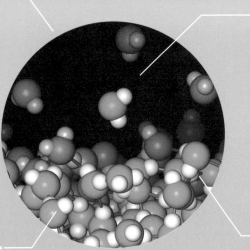

As the temperature of water increases, more and more molecules on the surface of the water have enough thermal energy to break the hydrogen bonds, and so evaporate to form a vapour. The evaporation becomes rapid as the temperature approaches 100 °C, and when the temperature reaches 100 °C, at sea level, the water boils

Water molecules have one oxygen atom (light blue) and two hydrogen atoms (white)

Each molecule is neutral, but the oxygen has a small negative charge and the hydrogens are slightly positive. The hydrogens are attracted to oxygens in nearby molecules, forming weak hydrogen bonds (dark blue)

1 SIMPLE KINETIC MOLECULAR MODEL OF MATTER

States of matter

Almost all matter can be classified as a solid, a liquid or a gas. These are called the three states of matter.

(The fourth state of matter is called 'plasma'. It only exists at high temperatures seldom seen on Earth, and so we won't consider it further here, even though most of the matter in the universe and most stars are made of plasma.)

As you will know, in general solids can be turned into liquids by heating, and with more heating liquids can be turned into gas. We are all familiar with water in all three states, but less so with other materials. Solid air is uncommon simply because it exists only at extremely low temperatures, and iron in the form of a gas only exists at very high temperatures.

The main body of this rocket is filled with liquid oxygen and liquid hydrogen, which have to be kept at extremely low temperatures to prevent them from heating up and turning back into gas. If the fuel were made colder it would turn into a solid.

The molten iron can be poured into a mould before it cools down and turns back into a solid.

All materials are made of tiny particles called atoms. The atoms attract each other, and the particles in a solid are locked together by the forces between them. But even in a solid the particles are not completely still. They vibrate constantly about their fixed positions. If the material is heated, it is given more internal energy, and the particles vibrate faster and further.

If the temperature is increased more, the vibrations of the particles increase to the point where the forces are no longer strong enough to hold the structure together. The forces are no longer enough to prevent the particles moving around, but they do prevent the particles from flying apart from each other. This is the liquid state. A liquid can flow and takes the shape of whatever container it is in. Its volume does not change much.

If the temperature is increased even more, then the particles do indeed fly apart. They now form a gas. The particles fly around at high speed (several hundred kilometres per hour) and if they are in a container, they travel all over the container, bouncing off the walls. The volume of a gas is not fixed, it just depends on the size of the container that the gas is put in.

	Solid	Liquid	Gas
Arrangement of particles	Regular pattern, closely packed together, particles held in place	Irregularly packed together, particles able to move past each other	Irregular, widely spaced, particles able to move freely
Diagram			
Motion of particles	Vibrate in place within the structure	'Slide' over each other in a random motion	Random motion, faster movement than the other states

Molecular model

The kinetic molecular model uses this idea that all materials are made up of atoms that behave rather like tiny balls. And with this idea we try to build up a simple explanation (a 'model') of as much as possible.

When the model is used to try to explain the behaviour of gases it is often called the kinetic theory of gases.

In the above diagram, you can see that the particles in the liquid and the gas consist of separate atoms. There are materials that behave like this, elements such as helium and neon. In most materials, though, the particles that move around in the liquid or the gas are groups of atoms called molecules. The water molecule is H_2O, and the nitrogen molecule is N_2. This means that the particles moving around in liquid or gaseous water each consist of two hydrogen atoms and one oxygen atom. And in liquid nitrogen or in nitrogen gas, the particles each consist of two nitrogen atoms.

Observed feature of a gas	Related ideas from the kinetic theory
Gases have a mass that can be measured.	The total mass of a gas is the sum of the masses of the individual molecules.
Gases have a temperature that can be measured.	The individual molecules are always moving. The faster they move (the more **kinetic energy** they have), the higher the temperature of the gas.
Gases have a pressure that can be measured.	When the molecules hit the walls of the container they exert a force on it. It is this force, divided by the surface area of the container, that we observe when measuring pressure.
Gases fill the volume of whatever container they are put in.	Although the volume of each molecule is only tiny, they are always moving about and spread out throughout the container.
Temperature has an absolute zero.	As temperature falls lower, the speed of the molecules (and their kinetic energy) becomes less. At absolute zero the molecules would have stopped moving.

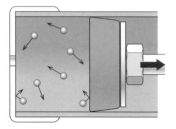

Let us see how these ideas can be used to explain the behaviour of gases.

THE EFFECT OF TEMPERATURE ON A GAS

The model says that the pressure on the walls of a container is caused by the collisions with the speeding molecules. You can feel this pressure if you try to hold a bicycle pump in the pushed-in position. (If the pump is faulty and allows the air to escape, this does not work.)

In the diagram, the piston is *not* moving. However, there is a force trying to push it out. It is clear that if the molecules travel faster then they will hit the piston in the pump more often and harder. The pressure on the piston and on the walls will go up. This is exactly what will happen if the air gets hotter.

Brownian motion

Evidence for the molecular model came from observations by Robert Brown in the early nineteenth century. When he looked through a high-magnification microscope at fine dust or pollen particles in water, he saw that the particles were all constantly moving in a jittery way, even though the water itself was stationary. Fine particles in the air do the same thing, even if the air is completely still.

The first full explanation was given by Einstein in 1905. As he explained, the movement is caused by these visible particles being hit from all sides by particles of water that are far smaller, and so are completely invisible. Each second, all sides of a dust particle are hit by millions of water particles. In general, opposite sides of the dust are hit by about the same number of water particles, so it does not move. But because the water particles are moving randomly, it is obviously possible that one side of the dust will suddenly be hit by a few more water particles than the other side. The resulting momentary force would have no detectable effect on a large object floating in the water, but it is enough to make the small dust particle move slightly. These tiny 'kicks' to the dust will come at random times and will move it in random directions.

The inner tube from a tyre has been pumped up with air before use as a toboggan. It is the pressure caused by the movement of the air molecules that keeps it inflated. Because the temperature is low, the inner tube will have needed more air. On a hot day this tube could burst.

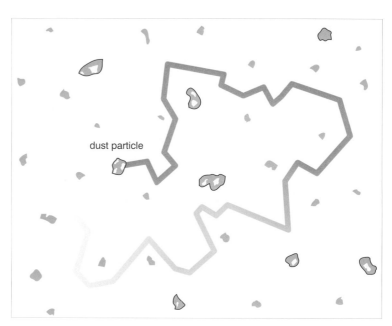

dust particle

Evaporation

When particles break away from the surface of a liquid and form a vapour, the process is known as evaporation. The more energetic molecules of the liquid escape from the surface. This reduces the average energy of the molecules remaining in the liquid and so the liquid cools down.

Evaporation causes cooling. The evaporation of sweat helps to keep a body cool in hot weather. The cooling obtained in a refrigerator is also due to evaporation of a special liquid inside the cooling panel at the back of the compartment.

Evaporation is increased at higher temperatures and it is also increased by a strong flow of air across the surface of the liquid, as in this way the evaporating molecules are carried away quickly. A certain amount of water will also evaporate more quickly if you increase its surface area. A bowl of soup will cool down much more quickly than a mug of soup, because the large surface area of the bowl allows more evaporation.

Clouds are formed from invisible water vapour that evaporates from the sea and is carried away by the wind. When the water vapour cools at high altitude, it turns back into the small droplets of water that you can see as these clouds.

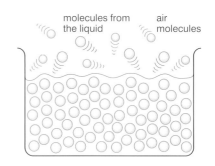

molecules from the liquid   air molecules

As water vapour cools, it forms clouds as seen here.

Pressure changes

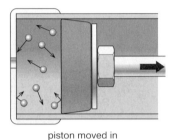

piston moved in

If the piston of the bicycle pump is pushed in, then the more that you push it in, the harder and harder it gets to push it further.

This is because the pressure in the container goes up. The molecular model says there are the same number of molecules in the container travelling at the same speed. However, because the molecules are now packed in more densely, there will be more collisions with the walls and with the piston per second. If the volume is halved, then the number of collisions with the walls and with the piston will double, and the pressure on the piston will double. This relationship between pressure and volume was discovered by Robert Boyle in 1662, and is often referred to as Boyle's law. It applies to any gas, but only if the temperature of the gas does not change during the measurements.

A fixed amount of gas in a sealed container at constant temperature obeys the following equation:

$$\text{pressure} \times \text{volume} = \text{constant}$$

$$pV = \text{constant}$$

p = pressure in Pa (or N/m^2 or millibar)

V = volume in m^3 (or cm^3)

Pascals and newtons per square metre are the same thing. Apart from them, you can use whichever units you like so long as you stick with them.

The constant will be a constant for a particular sample of gas in a particular container. So, in an experiment (or an exam question) you can write that the initial values of pressure and volume, $p_1 \times V_1 = \text{constant}$

And the final values of pressure and volume, $p_2 \times V_2 = \text{constant}$.

This is the same constant in both cases.

Hence $p_1 V_1 = \text{constant} = p_2 V_2$

or

$$p_1 V_1 = p_2 V_2$$

This equation, Boyle's law, only applies if the temperature stays constant.

It is important to emphasise that the temperature must remain constant. This is because the temperature of a gas will tend to go up if you compress it quickly. You may have noticed this: a bicycle pump gets very hot if you pump up a tyre quickly. So the law only applies if you measure the final pressure after the gas has cooled down again, or if you compress it very slowly so that its temperature does not change. By the same token, if you allow the gas to expand quickly, it cools down, so you have to take precautions here as well.

WORKED EXAMPLE

A bicycle pump contains 400 cm^3 of air at atmospheric pressure. If the air is compressed slowly, what is the pressure when the volume of the air is compressed to 125 cm^3? What happens to the pressure if the air is compressed quickly? (Remember that atmospheric pressure = 100 kPa.)

Write down the equation: $p_1V_1 = p_2V_2$

Substitute values into the equation: $100 \times 400 = p_2 \times 125$

$$p_2 \times 125 = 40\,000$$

Rearrange the equation to find p_2: $p_2 = \dfrac{40\,000}{125}$

Work out the answer and write down the unit: $= 320$ kPa

If the air is compressed quickly, it will also heat up to a higher temperature. This will mean that the final pressure will be greater than 320 kPa.

REVIEW QUESTIONS

Q1 Use ideas about particles to explain why:
 a solids keep their shape, but liquids and gases don't
 b solids and liquids have a fixed volume, but gases fill their container.

Q2 Use the kinetic molecular model to explain the following observations in detail:
 a It is possible to keep a bottle of drink cold by standing it in a bowl and covering it with a wet cloth.
 b The drink gets even colder if you place the bowl in a strong draught.
 c If two identical metal cylinders have most of the air removed from them, then they will balance when placed on a pair of scales. If one of the cylinders has air let back in, the scales will no longer balance.
 d An aerosol can of window cleaner has a large label on the side that says 'Danger. Do not dispose of the can by throwing it into a fire.'

Q3 A scuba diver has a 12 litre cylinder of air at 200 bar pressure. She is breathing the air at a depth of 20 m.
 a Calculate the volume of air available to this diver to breathe at this depth.
 b If each breath is 6 litres, and the diver is breathing 8 times per minute, how long can she stay at this depth? (In fact she would have to start ascending well before this time, as the ascent must be slow.)

More questions on the CD ROM

Examination questions are on page 82.

2 THERMAL PROPERTIES

Videos & questions
on the CD ROM

Thermal expansion of solids, liquids and gases

Why do things expand on heating?

With only two or three exceptions, all materials (solids, liquids and gases) expand as they become warmer. In the case of solids, the atoms vibrate more as the temperature goes up. So, even though they stay joined together, they move slightly further apart, and the solid expands a little in all directions.

The effect is small, but not trivial. A metre rule that is heated from 0 °C to 100 °C (from the freezing point of water to its boiling point) will get 1 to 2 mm longer depending what material it is made of. Some plastics would not make good metre rules, as they would get up to 10 mm longer.

On a hot day, a 1000 km railway track can try to become more than 300 m longer. In the early days of railways, there was a gap of a few mm every 20 m, to allow the rails to expand. Modern track has no expansion gaps, though as can be seen from this picture the track has to be held extremely firmly to stop it moving. This track is on a curve and so it will try to bend sideways to the left when it gets hot.

Liquids expand for the same reason. The atoms vibrate as they move around, and get slightly further apart. We talk about the increase in its volume as the temperature increases.

It is very difficult to prevent the thermal expansion of solids and liquids, as the material will create very large forces if it is not allowed to expand. So, for example a large bridge is always built with expansion joints to allow it to get longer on a hot day.

This thermometer has a bulb of coloured alcohol at the base, attached to a very narrow tube that is half-full of alcohol (ethanol). As the alcohol expands and contracts with change in temperature, the length of the column of alcohol goes up and down. The top of the tube is sealed off to prevent the alcohol from evaporating.

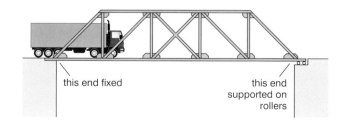

this end fixed

this end supported on rollers

62

However the effect can also be useful. Metals expand at different rates as their temperatures rise. So if strips of two metals are bound closely together, and are warmed, they bend as one metal expands more than the other. Bimetallic strips like this can be used to control the temperature in a heating system such as an electric iron.

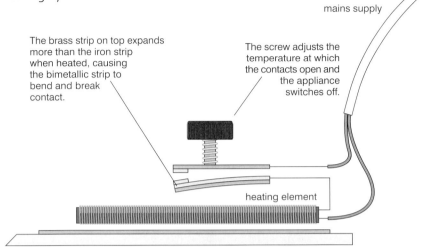

mains supply

The brass strip on top expands more than the iron strip when heated, causing the bimetallic strip to bend and break contact.

The screw adjusts the temperature at which the contacts open and the appliance switches off.

heating element

Like other liquids, water contracts as its temperature falls, and its density increases. Unlike other liquids, when its temperature falls below 4 °C, water begins to expand again, and becomes less dense. This is called the anomalous expansion of water. The density falls even further as it freezes, because the water molecules form an open crystal structure in the solid state. So ice is less dense than water, while almost all other materials are more dense in the solid state than as a liquid.

Gases behave completely differently. Firstly, we don't have to allow the gas to expand if it gets hotter; if we put it in a sealed container then we can just allow the pressure to increase instead. Secondly, if we do allow a gas to expand, then it will increase in volume much more than solids or liquids do as it gets hotter. Between 0 °C and 100 °C it will expand by a third, so 300 cm^3 of gas will become 400 cm^3.

In the diagram below the piston compresses the gas with a constant force so that the pressure of the gas is constant. We know from the molecular model that the piston is supported by the collisions of the molecules with the underside of the piston. If the temperature of the gas increases, the pressure starts to increase because the molecules travel faster, and they hit the piston harder. Hence the piston starts to move up. The piston stops moving up when the pressure has dropped to the original value.

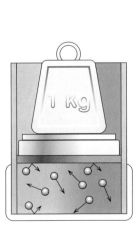

gas at low temperature

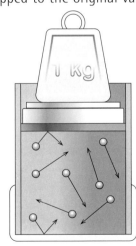

gas at high temperature

The result is that we have heated the gas and allowed its volume to increase at constant pressure. Note that initially the pressure was made up of many molecules hitting the piston slowly. After the gas has heated up and expanded, the same pressure is made of fewer collisions with the piston, but these collisions are from molecules moving faster.

Measurement of temperature

Temperature can be measured by any suitable physical property that changes with temperature. Examples in use include:
* volume of a liquid – mercury-in-glass or alcohol-in-glass thermometer
* length of a solid – bimetallic strip in a thermostat.

But many other properties are used:
* the pressure inside a fixed volume of gas
* the electrical resistance of a platinum resistor
* the electromotive force (e.m.f) generated by a thermocouple and so on.

All thermometers need calibrating before they are first used. In the case of a mercury-in-glass thermometer, this means that a scale must be fixed to it in the right place. To do this, two fixed points are required.

In science, the Celsius and Kelvin scales are used, though other scales are met elsewhere. The two fixed points used by the Celsius scale, as originally defined, are the melting point of ice, defined as 0 °C, and the boiling point of water at standard atmospheric pressure, defined as 100 °C. (At lower pressures it boils at a lower temperature.)

To calibrate the mercury-in-glass thermometer at these two fixed points, the thermometer is immersed first in a beaker containing melting ice, and the 0 °C point is marked. It is then immersed in the steam from a boiling kettle and the 100 °C point is marked. Finally the distance between the marks (distance *h* in the diagram) is divided into 100 equal distances, each corresponding to one degree. The scale can be extended beyond 100 °C to measure higher temperatures, and below 0 °C to measure negative temperatures.

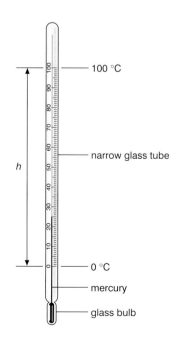

SENSITIVITY, RANGE AND LINEARITY

To measure temperature, we need a thermometer using a property that varies in a regular way over a suitable range of temperatures.

Sensitivity

A thermometer is **sensitive** if it gives a large response to a small change in temperature. This gives you a better chance to detect a small temperature change. To make a liquid-in-glass thermometer sensitive, you need a large bulb (so that the actual increase in volume is large) and a very narrow glass tube (so that the change in volume causes a large movement of the liquid up or down the tube). The use of a liquid that expands more than mercury is also helpful, and this is one way in which an alcohol-in-glass thermometer is better, because alcohol (ethanol) expands five times as much as mercury.

Range

Different thermometers have different ranges of temperatures that they can read. Mercury freezes at -38 °C, and so would be of no use in the Antarctic. Here one would need to use alcohol, which doesn't freeze until it gets down to -114 °C. Likewise, a thermometer filled with alcohol would be no use for measuring temperatures in an oven, as it would break when the temperature reaches 78 °C, and the liquid boils.

Linearity

When a thermometer is calibrated (see page 64), it is common to mark the ice point (0 °C) and the steam point (100 °C) and to divide the region in between into 100 equal parts. This assumes that the property doing the measuring changes by the same amount for every unit of temperature change. Such a property means that the thermometer has **linearity**. The expansion of the liquid in a liquid-in-glass thermometer does expand fairly linearly, so we can rely on the values we get from such thermometers.

Accuracy

An **accurate** thermometer is one that gives correct values of temperature. Students sometimes get confused between the terms **sensitive** and **accurate**. However, a sensitive thermometer is not necessarily an accurate one. A sensitive thermometer is one that can detect small changes in temperature, but if the scale has been incorrectly marked, or if the property doing the measuring varies in a noticeably non-linear manner, then the readings from this sensitive thermometer will not be accurate. Try to avoid using the word *accurate* in places where you mean *sensitive*.

THE THERMOCOUPLE

The thermocouple is an electrical thermometer that is the most common type used in industry. It is electrical, and so can be read on a remote dial; it can measure temperatures of over 1000 °C; and it is cheap to make. It is ideal therefore for measuring the temperature of inaccessible parts of a jet engine, and for measuring the temperature of a cauldron of molten steel.

Another major advantage of the thermocouple is that it can be made very small, which means that it will respond very quickly to a change in temperature; it can be made to respond in less than 1 second.

This oil refinery will use hundreds of thermocouples measuring the temperature at critical points, and sending the information to display panels and computers in the central control rooms.

The thermocouple is based on the fact that any two metals in contact generate a tiny voltage (actually a tiny e.m.f.). In order to measure this voltage with a voltmeter, the two metals need to form a circuit, which as in the diagram below, means that there must be two junctions. If the junctions are at the same temperature, there will be no voltage because the two voltages will cancel out, but if the junctions are at different temperatures, the difference between the two voltages can be measured with a voltmeter. One junction is placed at the point where the temperature is to be measured. The other junction is kept at room temperature, or for accurate work it is placed in a beaker of melting ice to take it to 0 °C.

There are tables of data available showing what voltage equals what temperature.

In practice, many different metals are used, but two metals often chosen are copper and an alloy called constantan. The junctions are made by twisting the wires tightly together.

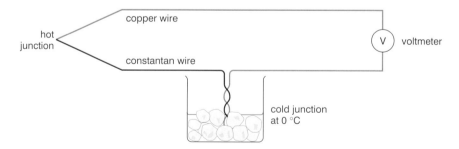

Thermal capacity

You make things hotter by transferring energy to them. When you heat up soup on a cooker ring, heat flows from the hot ring to the cold saucepan and then to the soup. You can also make things hot by rubbing or shaking them (see diagrams right). For example, rubbing your hands together helps to warm them up on a cold day. The kinetic energy of the movement is transferred to internal energy in your hands. A rise in temperature of an object shows that its internal energy has increased.

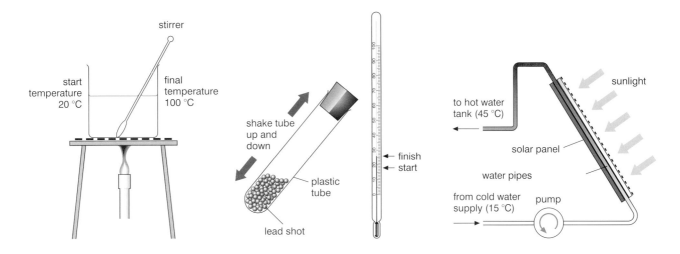

You need more energy to heat up a large amount of material than a small amount. All materials are made up of tiny particles. The larger the mass of the material, the more particles there are to share the added heat energy, so the smaller the temperature rise. Some materials are harder to heat up than others.

A large object, made of a material that takes a large amount of energy to heat up, will be able to store a great quantity of internal energy. We say that this object has a high thermal capacity. The thermal capacity of an object can be very important. It is the low thermal capacity of the thermocouple that means it can respond very quickly to changes in temperature. On the other hand it is the high thermal capacity of the world's oceans that allows countries near the sea to avoid suffering from extremes of temperature, as the oceans can release large amounts of heat if the land is cooler than the sea, and absorb it if the land is hotter.

This traditional mud house has very thick walls and is ideally suited for hot countries. The high thermal capacity of the walls causes the house to warm up slowly during the day and to cool down slowly at night.

The amount of energy (in joules) needed to raise the temperature of 1 kg of a material by 1 °C is called the specific heat capacity (see table below).

$$\text{specific heat capacity (J/kg °C)} = \frac{\text{energy used (J)}}{\text{mass (kg)} \times \text{temperature change (°C)}}$$

Material	Specific heat capacity/J/kg °C
Pure water	4200
Coal ash	900
Copper	390
Aluminium	910
Brick	800
Pyrex glass	780
Stainless steel	510
Concrete	3350
Magnetite (Fe_3O_4)	940

Specific heat capacities of different materials.

To measure the specific heat capacity of a solid material, drill two holes into a block of the material, one for a thermometer, and one for an electrical heater.

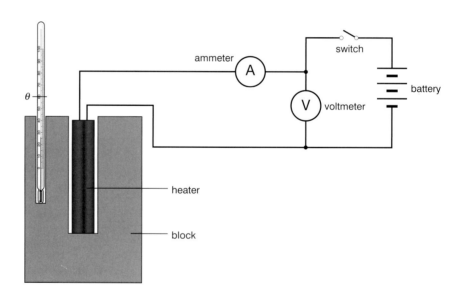

The electrical heater is connected to a low-voltage power supply or a battery. Two meters are required: an ammeter to measure the electrical current in the heater, and a voltmeter to measure the potential difference across the heater. As an alternative a joulemeter can be fitted in place of the ammeter.

The heater should have a start and stop switch, and you need an accurate thermometer to measure the temperature of the block. To make the experiment more accurate the surface of the block can be covered with thermal insulation such as expanded polystyrene to prevent heat loss.

The temperature of the block should be read, perhaps every 10 s. The heater should then be switched on for long enough to get a reasonable temperature rise.

The graph of results should look something like this:

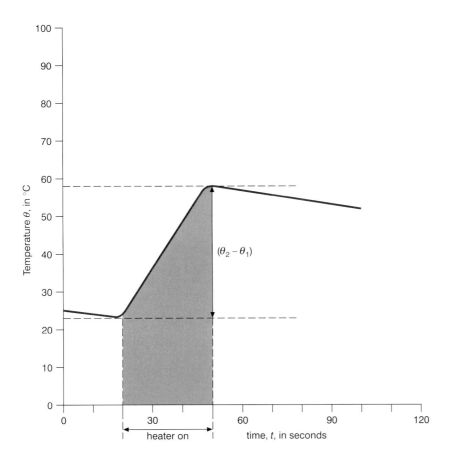

You will note that the maximum temperature is reached many seconds after the heater is switched off. This is because it takes time for the heat to spread through the block. From the graph the maximum temperature change ($\theta_2 - \theta_1$) can be measured.

In this graph, the block was still cooling down slowly when the experiment was started. It would have been better to wait for the temperature to become steady before starting.

Some of the physics here is covered in more detail in the section on electricity (see pages 112–141).

You can use the results of the experiment to calculate the specific heat capacity as shown on the next page.

The increase in internal energy E of the block is $\Delta E = V \times I \times t$

ΔE = change in internal energy in joules

V = potential difference in volts

I = electrical current in amps

t = time in seconds

Substituting in the values:

$$V = 24\,V$$
$$I = 7.5\,A$$
$$t = 20\,s$$

$$\Delta E = 24 \times 7.5 \times 20$$

Write down the answer and add the units: $\quad \Delta E = 3600\,J$

The increase in internal energy of the block is 3600 J

(Alternatively, with the joulemeter, you switch on the power for 20 s, and read the number of joules given to the block from the display of the meter.)

Now, the specific heat capacity of the material is $c = \dfrac{\Delta E}{m \times (\theta_2 - \theta_1)}$

m = mass of the block in kg

$\theta_2 - \theta_1$ = temperature change in °C

ΔE = change in internal energy, as above

c = specific heat capacity of the block in J/kg°C

Substituting in the values:

$$m = 0.2\,kg$$
$$\Delta E = 3600\,J$$
$$\theta_2 - \theta_1 = 45\,°C$$

$$c = \frac{3600}{(0.2 \times 45)}$$

$$= \frac{3600}{9}$$

Write down the answer and add the units: $\quad = 400\,J/kg°C$

The specific heat capacity of the block of material is 400 J/kg°C

WORKED EXAMPLE

An electric kettle has a power of 2 kW (2000 J/s). How long will it take to boil 1 l of water starting at 20 °C?

The specific heat capacity of water = 4200 J/kg°C .

Write down the equation: $\qquad c = \dfrac{\Delta E}{m \times (\theta_2 - \theta_1)}$

Rearrange the equation: $\qquad \Delta E = c \times m \times (\theta_2 - \theta_1)$

Now 1 l of water is 1 kg of water.

The water has to go from 20 °C to 100 °C, a rise of 80 °C.

Substitute the values: $\qquad \Delta E = 4200 \times 1 \times 80$

Work out the answer and write down the units: $\quad \Delta E = 336\,000\,J$

The water requires 336 000 J of heat to boil. The kettle can put 2000 J into the water each second. Therefore it will take 336 000/2000 seconds to boil = 168 s. Just under three minutes.

What about cooling down?

When the energy is being **transferred away** from an object, that is it is **cooling down**, the formula is used in exactly the same way. This time, instead of calculating the energy transferred to the object, the answer refers to the energy transferred away from it.

Melting and boiling

If you take a beaker that contains pieces of extremely cold ice and warm it up with an electrical heater, you can measure the temperature of the ice (and then water) every few seconds, and plot a graph of the temperature of the ice and water as the contents of the beaker warm up.

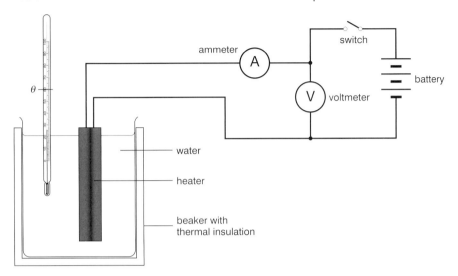

The graph will look like this:

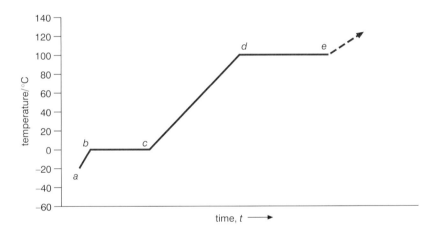

From point a to point b the ice is warming up, but it is not starting to melt. (This is similar to the behaviour of ice-cream after you take it out of the freezer, as it slowly takes heat out of the air and warms up without melting to begin with.)

Along line a–b, the heater is increasing the internal energy of the ice, and this is shown by the increase in temperature.

From point b to point c, the beaker contains a mixture of ice and water. The temperature stays constant, at 0 °C, and the heater melts the ice.

This makes a very important point. The energy from the heater has gone into the beaker, and so the internal energy of the contents of the beaker has gone up, but the temperature has not gone up. The energy has been used to melt the ice, and it is stored in the water. (In fact you have to put in almost as much energy to melt the ice as you will do, in the next step, to raise the water from freezing point to boiling point.) The same latent energy must be removed again to turn the water back into ice. (The word 'latent' means 'hidden'.) This is why it takes a freezer so long to freeze water.

From point c to point d, the input of heat energy into the water raises its temperature from 0 °C to 100 °C, and at point d the water boils.

In boiling, every particle in the liquid has enough energy to break away. This happens at a particular temperature – the **boiling point**. At the boiling point, the energy added to the material will be breaking the particles apart – the temperature does not change.

So from d to e the temperature of the boiling water stays constant, at 100 °C.

In this apparatus, the heater has to be turned off before all of the water is evaporated to prevent damage to the apparatus. But, in principle, if you could keep the steam, then after point e the heater could start to raise the temperature of the steam above 100 °C.

Condensation, the reverse of boiling, is where the gas turns into a liquid, and solidification is the reverse of melting.

Note that it is coincidence that we live in surroundings that are at a temperature of around 20–30 °C. That is why we see that steam tends to condense to water, and ice tends to melt to water. If we lived on a really hot planet like Mercury, then all water would tend to turn to steam. And on a cold planet, it would all tend to turn to ice.

LATENT HEAT

The extra energy stored by the water at 0 °C, as opposed to ice at 0 °C, is the **latent heat of fusion** (symbol L) of the water, and it is measured in joules. ('Fusion' is another word for solidifying.)

The heat required to melt a **unit mass** of solid, and turn it into liquid is known as the **specific latent heat of fusion** of that solid (symbol l), and is measured in J/kg or in J/g.

When you heat a liquid or a solid, and raise its temperature, the extra heat energy that you put in is stored as more vibration (in a solid) or more movement (in a liquid), as we have already discussed. Either way, it is definitely stored as a form of kinetic energy.

But while the solid is melting, the extra heat energy is used to weaken the bonds between the molecules and move the atoms slightly further apart against the attraction of the bonds. This energy is stored as potential energy, as in a stretched spring. So the internal energy of a liquid, or a gas, consists of some energy that is kinetic and some that is potential.

We use the term 'latent heat of vaporisation' to describe the energy that is needed to change the state from liquid to gas at the boiling point of the liquid.

The latent heat of vaporisation is the additional potential energy carried by the gas, stored in the broken bonds between the molecules.

This extra energy is carried by the steam, and is what makes steam so dangerous.

The heat required to turn a **unit mass** of liquid into gas is known as the **specific latent heat of vaporisation** (symbol l), and is measured in J/kg or in J/g.

AN EXPERIMENT TO MEASURE THE SPECIFIC LATENT HEAT OF FUSION OF ICE

To use the apparatus in the illustration (right), you must take a number of practical precautions both for safety and to make the experiment more accurate.

- You must fix the beaker of water so that the water cannot be spilled.
- You must fix lagging around the beaker to minimise the heat losses from the beaker. (See pages 77–81.)
- You may need to stir the water when the ice is nearly melted.
- You must connect a power supply to the heater with an on/off switch and with either a joulemeter or both a voltmeter and an ammeter so that you can calculate the energy being put into the beaker by the heater.

Using a heater to measure the latent heat of fusion for ice.

1. Weigh the beaker without the heater, and then assemble the apparatus.
2. Take some ice, and check that it is at 0 °C by noting that the outside of the ice is starting to melt.
3. Dry the ice on a cloth or tissues and put a suitable amount into the beaker.
4. Switch on the heater, and start a stopwatch. Note the readings on the meters.
5. When the ice is almost melted, start stirring the water gently.
6. Measure the time taken for the ice to melt.
7. Weigh the beaker plus the water.

WORKED EXAMPLE

Mass of beaker = 100 g

Mass of beaker + water = 237 g

Mass of water = 237 − 100
$\qquad\qquad$ = 137 g
$\qquad\qquad$ = 0.137 kg

The initial mass of ice was the same.

Time for ice to melt = 5 min 20 s
$\qquad\qquad\qquad$ = 320 s

Energy E put into the ice = $V \times I \times t$

E = change in internal energy in joules
V = potential difference in volts
I = electrical current in amps
t = time in seconds

Substitute the values: $\quad V = 24\,V$
$\qquad\qquad\qquad\qquad I = 6\,A$
$\qquad\qquad\qquad\qquad t = 320\,s$

$\qquad\qquad\qquad E = 24 \times 6 \times 320$

Work out the answer and write down the unit:
$\qquad\qquad\qquad = 46\,080\,J$

Specific latent heat of fusion $l = \dfrac{E}{m}$

E = change in internal energy in joules
m = mass in kg

Substitute the values: $\qquad\qquad\qquad l = \dfrac{46\,080}{0.137}$

Work out the answer and write down the unit: $\quad = 336\,350\,J\,kg$
$\qquad\qquad\qquad\qquad\qquad\qquad\qquad\qquad = 336\,kJ/kg$

This idealised experiment gives approximately the value that is generally accepted. If you try the same experiment you are likely to get a different reading, as it is all too easy to get a large error due to heat loss from the beaker and other factors that are hard to eliminate.

An alternative method is set as one of the questions at the end of this chapter (see page 76).

TO MEASURE THE SPECIFIC LATENT HEAT OF VAPORISATION OF WATER

1. Use the same apparatus as in the experiment on the previous page. Put sufficient water in the beaker and weigh the beaker plus water.
2. Put the heater into the beaker, and switch on. Check the readings on the meters.
3. When the water boils, start the stopwatch and let the water boil for a few minutes.
4. Let the beaker cool to a safe temperature and weigh it again.

WORKED EXAMPLE

Mass of beaker + water before the experiment = 237 g
Mass of beaker + water after the experiment = 218 g
Mass of water boiled off = (237 − 218)
$$= 19 \text{ g}$$
$$= 0.019 \text{ kg}$$

Water boiled for 5 minutes = 300 s

Energy E put into the boiling water	$= V \times I \times t$
Substitute the values:	$E = 24 \times 6 \times 300$
Work out the answer and write down the unit:	$E = 43\ 200 \text{ J}$
Specific latent heat of evaporation:	$I = \dfrac{E}{m}$
Substitute the values:	$I = \dfrac{43200}{0.019}$
Work out the answer and write down the unit:	$I = 2\ 273\ 684 \text{ J/kg}$
	$I = 2273 \text{ kJ/kg}$

Water is strange

Water has a surprisingly high specific heat capacity. This means that a lot of energy has to be transferred to change the temperature of water significantly. This is important in several ways.

- Water makes an excellent **coolant** for machines such as car engines. It can remove a lot of energy from the machine without boiling.

- The temperature of the seas and oceans remains fairly steady, as huge energy transfers are needed to significantly change the temperature of that much water. This helps keep the planet at a fairly even temperature, which is good for living things.

Note: A lot of confusion is caused by the different uses of the word 'steam'. In science, the word should be used to mean the invisible vapour that water turns into when it boils. This vapour is at 100 °C and is extremely dangerous to the human skin due to the energy that it contains. As soon as steam cools a little, it turns into the white clouds that we see when the kettle boils. These white clouds are made of small droplets of water, and are much safer than true steam. (It is true that in casual conversation we all call these clouds 'steam', but they are not steam in the scientific meaning.)

A* EXTRA

- The properties of water are very strange. Not only does it require a great deal of heat to change its temperature, it is also unique in that it expands as it freezes. This makes ice less dense than liquid water, so ice floats (as the *Titanic* found out). This has been vital to evolution – life can survive at the bottom of ponds, where the water stays liquid, even when the surface has frozen.

In this photograph, you can see that the steam coming out of the chimney is almost invisible. As the steam travels up, its temperature drops and it turns into clouds of water droplets.

The words 'evaporation' and 'boiling' also cause confusion. When a liquid **evaporates** it loses molecules from its surface. This will occur to an open container of water at any temperature; the water left in a cup will eventually evaporate. The molecules of water in the cup will have a range of energies. And even at room temperature the molecules with the highest energies will leave the surface.

A liquid **boils** when its temperature reaches boiling point. At this temperature the molecules have enough energy to leave the liquid in large quantities even inside the liquid. These molecules collect to form large bubbles of vapour and cause the liquid to move violently.

Many solids evaporate slowly, which is why you can smell dry food, coffee beans, for example. Some people even claim to be able to smell a sheet of zinc metal, even though the evaporation rate is incredibly low.

Note that the boiling point depends strongly on air pressure. People who live in high mountains have difficulty cooking food or making hot drinks, as the water may boil at 80 °C or lower, depending on the altitude. Astronauts wear spacesuits to prevent their blood from boiling at 37 °C.

REVIEW QUESTIONS

Q1 Explain the following observations:
  a A steel ruler is often marked 'Use at 20 °C'.
  b If you pour boiling water into a drinking glass, the glass may crack.
  c If you pour a very cold drink into a drinking glass, the outside of the glass will become wet.
  d If you leave frozen food in a freezer for several weeks without covering it, the outside surface of the food will suffer from what is called 'freezer burn' and will look dry and unpleasant.

Q2 A bimetallic strip consists of a thin strip of aluminium 100 mm x 10 mm attached to a thin strip of stainless steel of the same size.

aluminium

stainless steel

The two strips are joined together face to face, to give a thicker strip that is still 100 mm x 10mm. They are joined together at 20 °C using epoxy glue.

a The strip is fixed to a block of metal at one end. What happens to the other end at each change if the temperature goes to 100 °C, then to –10 °C, and finally back to 20 °C?

b Explain how this strip could be used as a thermometer.

c Design an electrical circuit that will use the bimetallic strip to switch on a warning light if the temperature of the bimetallic strip drops below room temperature and approaches the freezing point of water. Such a device is known as a 'frost stat' and is used to prevent damage from freezing.

d Explain what would happen if the warning light heated the bimetallic strip.

Q3 A student measures the specific latent heat of fusion of ice in the following way. She knows that the specific heat capacity of water is 4.2 J/g °C.

She takes a polystyrene beaker of negligible mass, and puts 400 g of warm water into it. She stirs the water gently, and measures its temperature with a thermocouple connected to a temperature display. The water temperature is 50 °C.

She then takes some wet ice that is at 0 °C, dries it on a tissue, drops it gently into the beaker and stirs until the ice has melted. The temperature is now 32 °C.

She checks the final weight of the beaker and finds that it is 460 g.

a The warm water cooled by how many degrees during the experiment? How many joules of heat did it give out?
The energy given out, $E = m \times c \times \Delta t$, where m is the mass of water in grams, c is the specific heat capacity in J/g °C and Δt is the temperature drop in °C.

b How many grams of ice did she add?

c After the ice had melted, it consisted of cold water at 0 °C. How many degrees did this then heat up during the experiment? How many joules of heat did this cold water take in as it heated up?

d How many joules of heat are still unaccounted for?

e If all of these joules of heat were used to melt the ice, what answer does she get for the specific latent heat of the ice in J/g?

Examination questions are on page 82.

More questions
on the CD ROM

3 TRANSFER OF THERMAL ENERGY

Energy will always try to flow from areas at high temperatures to areas at low temperatures. This is called **thermal transfer**. Thermal energy can be transferred in three main ways:

- **conduction**
- **convection**
- **radiation**.

Videos & questions on the CD ROM

Conduction

Materials that allow thermal energy to transfer through them quickly are called **thermal conductors**. Those that do not are called **thermal insulators**. (If someone talks about an 'insulator', you may have to work out for yourself if he is referring to a thermal insulator or to an electrical insulator.)

If one end of a conductor is heated, the atoms that make up its structure start to vibrate more vigorously. As the atoms in a solid are linked together by chemical bonds, the increased vibration can be passed on to other atoms. The energy of movement (kinetic energy) passes through the whole material.

Metals are particularly good thermal conductors because they contain freely moving electrons which transfer energy very rapidly. As the electrons travel all over the piece of metal, they take the thermal energy with them. This is in addition to the thermal energy that is transferred by vibrations of the atoms that make up the structure of the metal.

Conduction cannot occur when there are no particles present, so a vacuum is a perfect insulator.

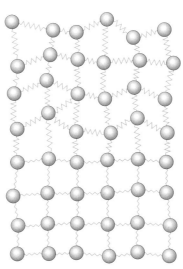

Conduction in a solid. Particles in a hot part of a solid (top) vibrate further and faster than particles in a cold part (bottom). The vibrations are passed on through the bonds from particle to particle.

EXPERIMENT TO SHOW CONDUCTION

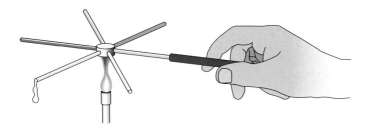

The rods are made of different metals, so the heat conducts along them at different rates. The better the conductor, the quicker the wax at the end of the rod melts.

Convection

Convection occurs in **liquids** and **gases** because these materials flow (they are 'fluids'). The particles in a fluid move all the time. When a fluid is heated, energy is transferred to the particles, causing them to move faster and further apart. This makes the heated fluid less dense than the unheated fluid. The less dense warm fluid will rise above the more dense colder fluid, causing the fluid to circulate. This **convection current** is how the thermal energy is transferred.

If a fluid's movement is restricted, then energy cannot be transferred by convection. That is why many insulators, such as ceiling tiles, contain trapped air pockets. Wall cavities in houses are filled with fibre to prevent air from circulating and transferring thermal energy by convection.

EXPERIMENT TO SHOW CONVECTION

Potassium permanganate crystals in water demonstrate convection. The warmer water expands, becomes less dense and rises, making a trail as some of the dissolved potassium permanganate is carried along as well.

Radiation

Radiation, unlike conduction and convection, **does not need particles** at all. Radiation can travel through a vacuum. This is clearly shown by the radiation that arrives from the Sun. Radiated heat energy is carried mainly by infrared radiation, which is part of the electromagnetic spectrum.

All objects take in and give out infrared radiation all the time. Hot objects radiate more infrared than cold objects. The amount of radiation given out or absorbed by an object depends on its temperature and on its surface.

Type of surface	As an emitter of radiation	As an absorber of radiation	Examples
Dull black	Good	Good	Emitter: Cooling fans on the back of a refrigerator are dull black to radiate away more energy. Absorber: The surface of a black bitumen road gets far hotter on a sunny day than the surface of a white concrete one.
Bright shiny	Poor	Poor	Emitter: Marathon runners, at the end of a race, wrap themselves in shiny blankets to prevent them from cooling down too quickly by radiation (or convection). Absorber: Fuel storage tanks are sprayed with shiny silver or white paint to reflect radiation from the Sun.

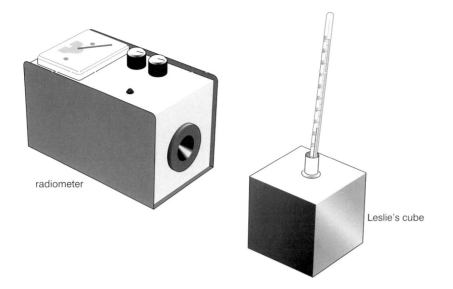

radiometer

Leslie's cube

A radiometer is an instrument that allows you to measure the heat being radiated by a surface. If you want to compare the heat radiated by different surfaces, then Leslie's cube allows you to compare the effect of four different surfaces. These four surfaces, which might, for example, be shiny metal and dull metal, white paint and black paint, are heated to the same temperature by hot water inside. Hence any difference in the heat radiated is entirely due to the nature of the surface being measured.

Consequences of energy transfer

THE RADIATOR

A radiator does radiate some heat, and if you stand near a hot radiator your hands can feel the infrared radiation being emitted by the front surface of the radiator. However, this is only around one quarter of the heat being released by the radiator. *Three quarters* of the heat is taken away by the hot air that rises from the radiator. Colder air from the room flows in to replace this hot air, and a convection current is formed as shown.

This shows a side view of a room with a hot-water radiator underneath the window.

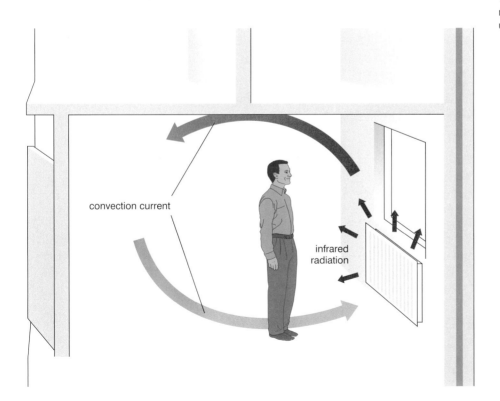

convection current

infrared radiation

You will note that the convection current is far more efficient at heating the top of the room than it is at heating the person standing in front of the radiator.

THE VACUUM FLASK

The vacuum flask will keep a drink hot or cold for hours by almost completely preventing the flow of heat out or in.

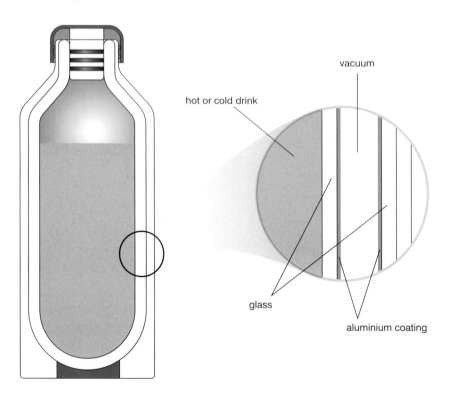

Conduction is almost eliminated by making sure that any heat flowing out must travel along the glass of the neck of the flask. The path is a long one, the glass is thin, and glass is a very poor conductor of heat. The bung in the top of the flask must also be a very poor conductor of heat: cork or expanded polystyrene is good.

Convection is eliminated because the space between the inner wall and the outer wall of the flask is made a vacuum so that there is no air to form convection currents.

If the contents are hot, radiation is almost eliminated because the inner walls of the flask are coated with pure aluminium. Because the aluminium is in a vacuum, it stays extremely shiny forever, and so the wall in contact with the hot liquid emits very little infrared radiation.

The vacuum flask works just as well the other way, at keeping the contents cool. And in fact it was invented in the 1890s by a Scottish physicist so that he could keep liquefied gases, such as liquid hydrogen that boils at −250 °C. Scientists usually call it a Dewar flask in his honour.

EFFICIENCY OF ENERGY TRANSFER

Energy transfers can be summarised using simple **energy transfer diagrams** or **Sankey diagrams**. The thickness of each arrow is drawn to scale to show the amount of energy.

Energy is always conserved – the total amount of energy after the transfer must be the same as the total amount of energy before the transfer. Unfortunately, in nearly all energy transfers some of the energy will end up as 'useless' heat.

In a power station only some of the energy originally produced from the fuel is transferred to useful electrical output. Energy **efficiency** can be calculated from the following formula:

$$efficiency = \frac{useful\ energy\ output}{energy\ input} \times 100\%$$

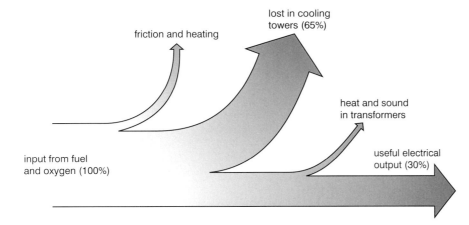

friction and heating

lost in cooling towers (65%)

heat and sound in transformers

input from fuel and oxygen (100%)

useful electrical output (30%)

In a power station as much as 70% of the energy transfers do not produce useful energy. The power station is only 30% efficient.

Many power stations are now trying to make use of the large amounts of energy 'lost' in the cooling system. In some cities, the houses of whole regions of the city are heated by hot water from the power station.

REVIEW QUESTIONS

Q1 Why are several thin layers of clothing more likely to reduce thermal transfer than one thick layer of clothing?

Q2 A student sets up an experiment. She places three shallow dishes each containing 1 cm of water on the ground. Dish A is in the shade and out of any draught; dish B is in the light from the sun; and dish C is both in the light from the sun and it is exposed to a strong wind.
  a She measures the levels in the three dishes every hour. What will she observe?
  b Explain her observations using the molecular model.
  c She measures the temperature of dish B, and finds that it goes up after all of the water has disappeared. Explain why.

Q3 The diagram shows a cross-section of a steel radiator positioned in a room next to a wall.

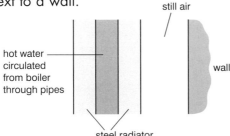

still air

hot water circulated from boiler through pipes

wall

steel radiator

Describe how energy from the hot water reaches the wall behind the radiator.

More questions on the CD ROM

Examination questions are on page 82.

EXAMINATION QUESTIONS

Q1   Fig. 1.1 shows a thermocouple set up to measure the temperature at a point on a solar panel.

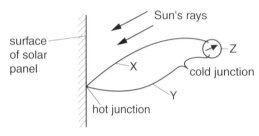

Fig. 1.1

a   X is a copper wire.
   i   Suggest a material for Y.

   _____

   ii   Name the component Z.

   _____ [2]

b   Explain how a thermocouple is used to measure temperature.

   _____

   _____

   _____ [3]

c   Experiment shows that the temperature of the surface depends upon the type of surface used.
   Describe the nature of the surface that will cause the temperature to rise most.

   _____

   _____ [1]

Q2   a   Fig. 2.1 shows a cylinder containing air at a pressure of 1.0×10^5 Pa. The length of the air column in the cylinder is 80 mm.

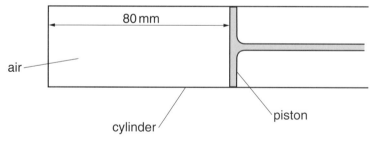

Fig. 2.1

The piston is pushed in until the pressure in the cylinder rises to 3.8×10^5 Pa.
Calculate the new length of the air column in the cylinder, assuming that the temperature of the air has not changed.
new length = _____ [3]

b   Fig. 2.2 shows the same cylinder containing air.

Fig. 2.2

The volume of the air in the cylinder changes as the temperature of the air changes.
   i   The apparatus is to be used as a thermometer. Describe how two fixed points, 0 °C and 100 °C, and a temperature scale could be marked on the apparatus.

ii   Describe how this apparatus could be used to indicate the temperature of a large beaker of water.

_____

_____

_____

_____ [5]

Q3 Fig. 3.1 shows water being heated by an electrical heater. The water in the can is not boiling, but some is evaporating.

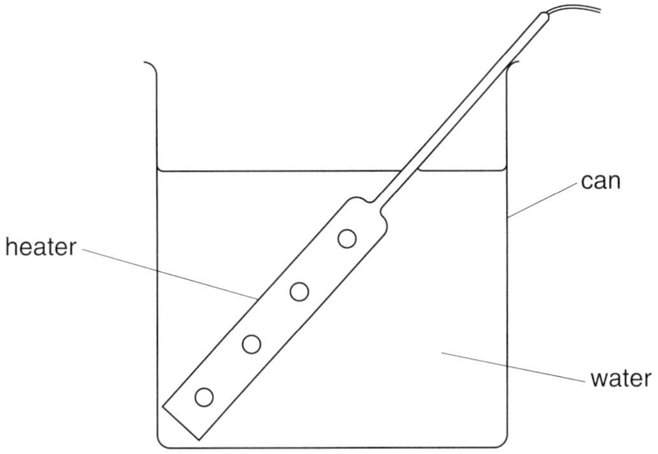

Fig. 3.1

a   Describe, in terms of the movement and energies of the water molecules, how evaporation takes place.

_____

_____

_____

_____ [2]

b   State two differences between evaporation and boiling.

1 _____

_____

2 _____ [2]

c   After the water has reached its boiling point, the mass of water in the can is reduced by 3.2 g in 120 s. The heater supplies energy to the water at a rate of 60 W. Use this information to calculate the specific latent heat of vaporisation of water.

_____

_____

_____

specific latent heat = _____ [3]

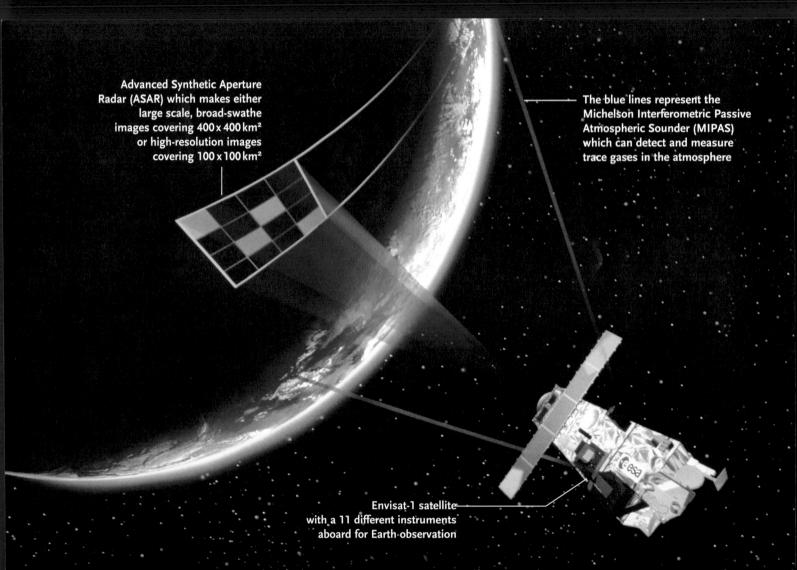

Advanced Synthetic Aperture Radar (ASAR) which makes either large scale, broad-swathe images covering 400 x 400 km² or high-resolution images covering 100 x 100 km²

The blue lines represent the Michelson Interferometric Passive Atmospheric Sounder (MIPAS) which can detect and measure trace gases in the atmosphere

Envisat-1 satellite with a 11 different instruments aboard for Earth observation

Eyes in the sky

You probably know that artificial satellites orbit the Earth. You can probably describe some ways in which the satellites are used. But you might be surprised at just how many uses there are. Did you realise that some newspapers send their text and pictures to different printing sites via satellite? This saves on distribution time and costs and allows the newspaper to keep as up to date with news stories as possible. You are probably aware of the Global Positioning System (GPS) that allows ships and planes to know their exact position on the Earth's surface, but did you realise the same system is used to track packages being delivered by trucks and is even used to tell when the drivers are driving too fast?

Satellite images use many different wavelengths of radiation, for example, visible wavelengths and infrared. They show different features of the surface beneath them, for example, some wavelengths are reflected by clouds and are useful in weather forecasting. Next time you see a weather forecast showing satellites images taken during the night, see if you can work out how the picture was taken. Wasn't it dark at the time?

There are many, many other applications of satellite imaging, from spying to search and rescue, from surveying to earthquake monitoring. Satellites come in different sizes and orbit the Earth in different paths. Why not spend a little time researching this interesting and constantly changing topic?

PROPERTIES OF WAVES, INCLUDING LIGHT AND SOUND

Satellite image of Singapore and surroundings
taken by NASA's *Landsat 7*
combining infrared and visible light

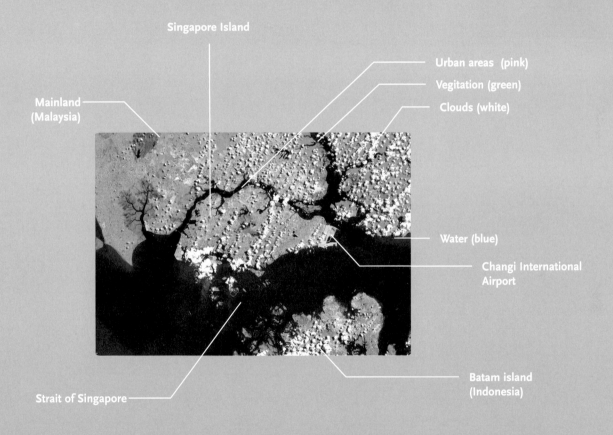

Singapore Island

Urban areas (pink)

Vegitation (green)

Clouds (white)

Mainland
(Malaysia)

Water (blue)

Changi International
Airport

Strait of Singapore

Batam island
(Indonesia)

1 GENERAL WAVE PROPERTIES

This student can feel the heat waves from the Sun coming in through the windows of the train, she can hear the sound waves of her friend on the phone, the phone is using radio waves, and she can see around her with light waves.

The behaviour of waves affects us every second of our lives. Waves are reaching us constantly: sound waves, light waves, infrared heat, television, mobile-phone and radio waves, the list goes on. The study of waves is, perhaps, truly the central subject of physics.

There are two types of waves: longitudinal and transverse.

Longitudinal waves. This type of wave can be shown by pushing and pulling a spring. The vibrations of the spring as the wave goes past are backwards and forwards in the direction that the wave is travelling (hence the name 'longitudinal'). The wave consists of stretched and squashed regions travelling along. The stretching produces regions of rarefaction, while the squashing produces regions of compression. Sound is an example of a longitudinal wave.

Transverse waves. In a transverse wave the vibrations are at right angles to the direction of motion. Light, radio and other electromagnetic waves are transverse waves.

In the above examples, the waves are very narrow, and are confined to the spring or the string that they are travelling down. Most waves are not confined in this way. Clearly a single wave on the sea, for example, can be hundreds of metres wide as it moves along.

Water waves are often used to demonstrate the properties of waves because the **wavefront** of a water wave is easy to see. A wavefront is the moving line that joins all the points on the crest of a wave.

Longitudinal and transverse waves are made by vibrations.

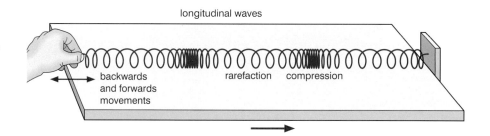

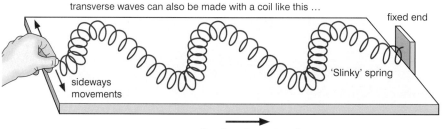

WHAT FEATURES DO ALL WAVES HAVE?

The speed a wave travels at depends on the substance or medium it is passing through.

Waves have a repeating shape or pattern.

Waves carry energy without moving material along.

Waves have a wavelength, frequency, amplitude and time period.

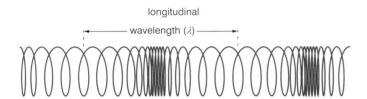

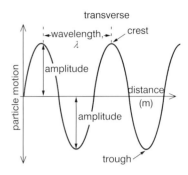

The **wavelength** is the distance between two adjacent peaks or, if you prefer, the distance between two adjacent troughs of the wave. In the case of longitudinal waves, it is the distance between two points of maximum compression, or the distance between two points of minimum compression.

The **frequency** is the number of peaks (or the number of troughs) that go past each second.

The **amplitude** is the maximum particle displacement of the medium from the central position. In transverse waves, this is half the crest-to-trough height.

The **speed** of the wave is simply the speed of the wave as it approaches a ship. The largest ocean wave ever measured accurately had a wavelength of 340 m, a frequency of 0.067 Hz (that is to say one peak every 15 s), and a speed of 23 m/s. The amplitude of the wave was 17 m, so the ship was going 17 m above the level of a smooth sea and then 17 m below. (The waves were 34 m from crest to trough.)

The **period** (T) is the time taken for each complete cycle of the wave motion. It is closely linked to the frequency (f) by this relationship:

$$\text{frequency (in hertz, Hz)} = \frac{1}{\text{period (in seconds, s)}}$$

The speed of a wave in a given medium is constant. If you change the wavelength, the frequency *must* change as well. If you imagine that some waves are going past you on a spring or on a rope, then they will be going at a constant speed. If the waves get closer together, then more waves must go past you each second, and that means that the frequency has gone up. The speed, frequency and wavelength of a wave are related by the equation:

wave speed = frequency × wavelength

$v = f \times \lambda$

v = wave speed, usually measured in metres/second (m/s)

f = frequency, measured in cycles per second or hertz (Hz)

λ = wavelength, usually measured in metres (m)

WORKED EXAMPLES

1 A loudspeaker makes sound waves with a frequency of 300 Hz. The waves have a wavelength of 1.13 m. Calculate the speed of the sound waves.

Write down the formula:	$v = f \times \lambda$
Substitute the values for f and λ:	$v = 300 \times 1.13$
Work out the answer and write down the unit:	$v = 339$ m/s

2 A radio station broadcasts on a wavelength of 250 m. The speed of the radio waves is 3×10^8 m/s. Calculate the frequency.

Write down the formula with f as the subject:	$f = \dfrac{v}{\lambda}$
Substitute the values for v and λ:	$f = \dfrac{3 \times 10^8}{250}$
Work out the answer and write down the unit:	$f = 1\,200\,000$ Hz or 1200 kHz

3 A tuning fork is used to play a middle C, which has a frequency of 256 Hz. Calculate the time period of the vibration.

Write down the formula with T as the subject:	$T = \dfrac{1}{f}$
Substitute the value of f:	$T = \dfrac{1}{256}$
Work out the answer and write down the unit:	$T = 0.0039$ s

REFLECTION, REFRACTION AND DIFFRACTION

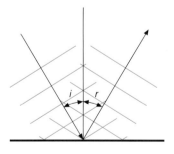

Waves hit a barrier at an angle of incidence i. The waves bounce off with the angle of incidence i equal to the angle of reflection r. The reflected wave is the same shape as the incident wave.

When a wave hits a barrier the wave will be **reflected**. If it hits the barrier at an angle then the **angle of reflection** will be **equal** to the **angle of incidence**. **Echoes** are caused by the reflection of sound waves.

When a wave moves from one medium into another, it will either speed up or slow down. For example, a wave going along a rope will speed up if the rope becomes thinner. (This is why you can 'crack' a whip: the wave that is sent down the whip accelerates until it breaks the sound barrier.) And sound going from cold air to hotter air will speed up. When a wave **slows down**, the wavefronts crowd together – the **wavelength gets smaller**. When a wave **speeds up**, the wavefronts spread out – the **wavelength gets larger**. Note that in both cases, the same number of waves will pass you per second, the wavelength may have changed, but the frequency has not.

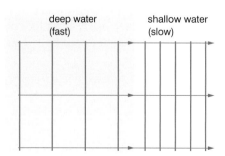

deep water (fast) shallow water (slow)

When waves slow down, their wavelength gets shorter.

This surfer is successfully travelling along one wavefront. The next wavefront looks very close behind, but is probably still 50 m away. (This is an illusion caused by telephoto camera lenses.)

If a wave enters a new medium at an angle then the wavefronts also change direction. This is known as **refraction**. The amount that the wave is bent by depends on the change in speed. Water waves are slower in shallower water than in deep water, so water waves will refract when the depth changes.

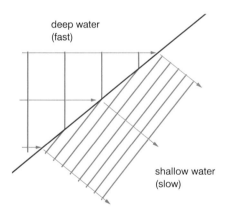

deep water
(fast)

shallow water
(slow)

If waves cross into a new medium at an angle, their wavelength and direction change.

Wavefronts change shape when they pass the edge of an obstacle or go through a gap. This process is known as **diffraction**. Diffraction is strong when the width of the gap is similar in size to the wavelength of the waves.

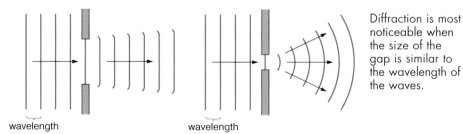

wavelength

wavelength

Diffraction is most noticeable when the size of the gap is similar to the wavelength of the waves.

In the first diagram the wavelength is much smaller than the aperture. An example is light coming in through a window and forming a beam of light across the room. In the second diagram, the wavelength is similar to the size of the aperture. Ocean waves spread out like this when they enter a harbour.

A* EXTRA

- High notes from a CD track have wavelengths of around 10–20 cm. For these to diffract efficiently and spread out as they leave the opening of the loudspeaker, a speaker with a smaller diameter is used. For lower notes, which have larger wavelengths, a speaker with a larger diameter is used in order to generate enough volume.

If the apertures are much smaller than the wavelength then the wave cannot go through the aperture at all, which is why the door of a microwave oven has an array of small holes for you to see inside. Light can travel in or out through the holes, but microwaves (with a wavelength of 12 cm) cannot do so.

Short-wavelength signals with a wavelength of less than 1 m are used by television and mobile phone transmitters. These wavelengths are much smaller than the size of buildings, so the waves do not bend round the buildings, and you will not get a good television signal or mobile phone reception if there is a building between you and the transmitter.

To reduce this problem for mobile phones, the mobile phone companies use many transmitters in a city, and switch your phone to the transmitter with the best path to the phone. Satellite television does not have a problem because the receiver on the house has a direct view of the satellite.

Long-wavelength radio signals, with a wavelength of 100 m or so, bend round buildings and hills, and can give a good reception anywhere in a city and even hundreds of kilometres from the transmitter.

REVIEW QUESTIONS

Q1 The diagram shows a graph of displacement against distance.

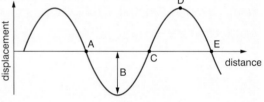

  a Which letter shows the crest of the wave?
  b The wavelength is the distance between which two letters?
  c Which letter shows the amplitude?
  d The frequency of the wave is 512 Hz. How many waves are produced each second?

Q2 Radio waves of frequency 900 MHz are used to send information to and from a portable phone. The speed of the waves is 3×10^8 m/s. Calculate the wavelength of the waves. (1 MHz = 1 000 000 Hz, 3×10^8 = 300 000 000.)

Q3 What are the most likely explanations of the following effects? Explain carefully.
  a The captain of an ocean-going ship is proceeding slowly into waves that are coming towards the ship. He suddenly notices that the waves change in two ways about 200 m ahead of where the ship is. They get further apart and change direction quite noticeably.
  b An observer is standing on the bank of a river. The wind is blowing from left to right, and waves are moving from left to right. However, the observer sees a small piece of wood floating in the middle of the river that is moving slowly from right to left.
  c You find that you can listen to radio stations in all of the rooms in your home, but you cannot get a mobile phone signal in certain rooms even if you open the windows.

More questions on the CD ROM

Examination questions are on page 104.

2 LIGHT

Reflection of light

A ray of light is a line drawn to show the path that the lightwaves take.

We need to study what happens when an incident light ray (a light ray that is going to fall on a surface) hits a mirror and is reflected off.

Light rays are reflected from mirrors in such a way that

angle of incidence (i) = angle of reflection (r)

The angles are measured to an imaginary line at 90° to the surface of the mirror. This line is called the **normal**. With a curved mirror it is difficult to measure the angle between the ray and the mirror, but the same law still applies.

When you look in a plane mirror you see an **image** of yourself. The image is said to be **laterally inverted** because if you raise your right hand your image appears to raise its left hand. The image appears to be as far behind the mirror as you are in front of it and is the same size as you. The image cannot be projected onto a screen. It is known as a **virtual image**.

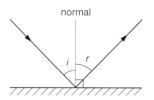

The angles of incidence and reflection are the same when a mirror reflects light. This type of curved mirror is known as a convex mirror.

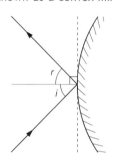

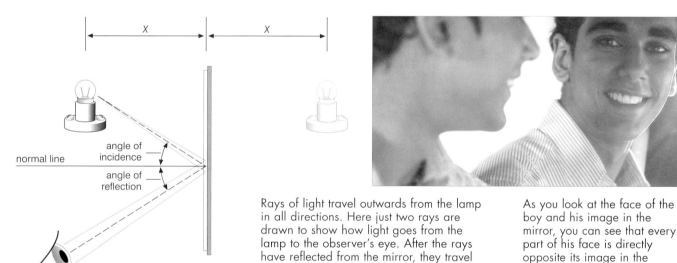

Rays of light travel outwards from the lamp in all directions. Here just two rays are drawn to show how light goes from the lamp to the observer's eye. After the rays have reflected from the mirror, they travel along lines that look *as if* they started from the image. The brain is tricked into thinking that the light really did start from the image.

As you look at the face of the boy and his image in the mirror, you can see that every part of his face is directly opposite its image in the mirror, and that each part is the same distance away from the mirror as its image.

Summary

An image formed by a plane mirror, such as the one above, is:
- virtual
- laterally inverted
- the same size as the object
- the same distance behind the mirror as the object is in front of the mirror.

Note that the image is never formed on the surface of the mirror. This is a mistake which is often made by candidates in examinations.

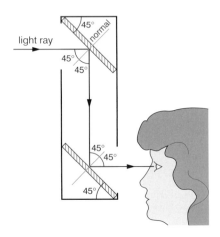

A periscope uses reflection to allow you to see above your normal line of vision – or even round corners.

USES OF MIRRORS

In a plane mirror the image is always the same size as the object. Examples of plane mirrors include household 'dressing' mirrors, dental mirrors for examining teeth, security mirrors for checking under vehicles, periscope.

Close to a **concave mirror** the image is **larger** than the object – these mirrors **magnify**. They are used in make-up and shaving mirrors. They are used in torches and car headlamps to produce a beam of light.

The image in a **convex mirror** is always **smaller** than the object. Examples are a car driving mirror and a shop security mirror.

Refraction of light

Light waves **slow down** when they travel from air into glass. If they are at an angle to the glass, they bend towards the normal. When the light rays travel out of the glass into the air, their speed increases and they bend away from the normal. If the block of glass has parallel sides, the light resumes its original direction. This is why a sheet of window glass has so little effect on the view beyond. However, the view is shifted slightly sideways if you are looking through the glass at an angle.

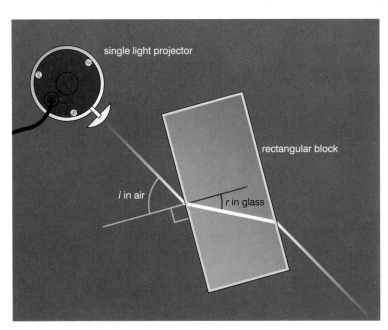

The angle of incidence i, is the angle between the incident light ray and the normal to the surface. The angle of refraction r is the angle between the refracted light ray and the normal to the surface inside the material.

(Note that people tend to use the letter r both for the angle of reflection and the angle of refraction. But it will be clear from the context whether they are talking about reflection or refraction.)

The **refractive index** of a material indicates how strongly the material changes the direction of the light. It is calculated using the following formula:

$$\text{refractive index}, n = \frac{\sin i}{\sin r}$$

The refractive index of a vacuum is 1, and the refractive index of air is fractionally higher, but we will take it as 1. Other common refractive indices are water 1.3; window glass 1.5; sapphire 1.75; diamond 2.4.

The refractive index n can also be defined as:

$$n = \frac{\text{speed of light in vacuum (or air)}}{\text{speed of light in the material}}$$

WORKED EXAMPLES

1 If the speed of light in a vacuum is 300 000 000 m/s, what is the speed of light in glass with a refractive index of 1.5?

Write down the formula: $n = \dfrac{\text{speed of light in vacuum (or air)}}{\text{speed of light in the material}}$

Rearrange the formula: $\text{speed in material} = \dfrac{\text{speed in vacuum}}{n}$

Substitute the values: $\text{speed in material} = \dfrac{300\ 000\ 000}{1.5}$

Work out the answer and write down the unit:
$$\text{speed of light in the material} = 200\ 000\ 000 \text{ m/s.}$$
$$= 2 \times 10^8 \text{ m/s}$$

2 A light ray approaches a block of plastic with an angle of incidence of 60°. If the refractive index of the plastic is 1.4, what is the angle of refraction?

Write down the formula: $n = \dfrac{\sin i}{\sin r}$

If $i = 60°$, $\sin i = 0.866$

Rearrange the formula: $\sin r = \dfrac{\sin i}{n}$

Substitute the values: $\sin r = \dfrac{0.866}{1.4}$

Work out the answer: $\sin r = 0.619$

From a calculator, if $\sin r = 0.619$, then $r = 38.2°$

The angle of refraction is 38.2°.

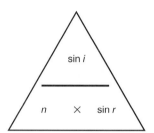

Total internal reflection

When rays of light pass from a **slow medium** to a **faster medium** they move **away** from the normal.

As the angle of incidence increases, an angle is reached at which the light rays will have to leave with an angle of refraction greater than 90°! These rays cannot refract, so they are entirely reflected back inside the medium. This process is known as **total internal reflection**.

The angle of incidence at which all refraction stops is known as the **critical angle** for the material. The critical angle of window glass is 42°, the critical angle of water is 49°.

The critical angle, c, is linked to the refractive index by this formula:

$$\sin c = \frac{1}{n}$$

Total internal reflection occurs when a ray of light tries to leave the glass at an angle of refraction greater than 90°. If the angle of incidence equals or is greater than the critical angle the ray will be totally internally reflected. In this picture, the angle is 2–3° less than the critical angle, and light is just managing to escape from the glass.

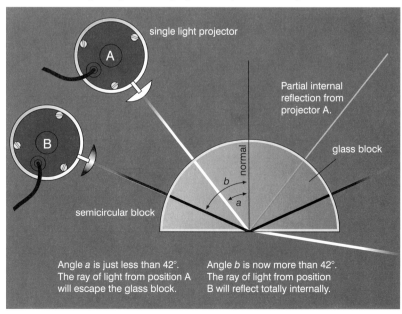

Angle *a* is just less than 42°. The ray of light from position A will escape the glass block.

Angle *b* is now more than 42°. The ray of light from position B will reflect totally internally.

Light does not escape from the fibre because it always hits it at an angle greater than the critical angle and is internally reflected.

Total internal reflection is used in **fibre–optic cables**. A fibre-optic cable is made of a bundle of very thin glass fibres. The light continues along the fibre by being constantly internally reflected.

Telephone and TV communications systems are increasingly relying on fibre optics instead of the more traditional copper cables. Fibre-optic cables do not use electricity and the signals are carried by infrared rays. The signals are very clear as they do not suffer from electrical interference. Other advantages are that they are cheaper than the copper cables and can carry thousands of different signals down the same fibre at the same time.

Bundles of several thousand optical fibres are use in medical endoscopes for internal examination of the body. The bundle will carry an image from one end of the bundle to the other, each fibre carrying one tiny part (one pixel) of the image.

Thin converging lens

Convex (converging, positive) lenses cause parallel rays of light to **converge**. As you can see, the light rays are bent by refraction, as described on page 89, except that the amount of refraction increases from the middle of the lens (where there is no refraction) to a maximum at the perimeter of the lens. The point where the light rays all cross over is known as the **principal focus**, and the **focal length *f*** is the most important feature of the lens. Note that there are two principal foci. (The word 'foci' is the plural of 'focus'.) The foci are each side of the lens, at the same distance away from it.

The principal axis is the line through the middle of the lens and at right angles to it. The principal foci lie on this line.

Converging lenses are used to form **images** in magnifying glasses, cameras, telescopes, binoculars, microscopes, film projectors and spectacles for long-sighted people.

You can check your glasses to see if they act as a weak magnifying glass, or if they make things seem smaller. If the glasses magnify, then you are long-sighted and you see distant things more clearly. If they do the opposite, then you are short-sighted and see close-up things clearly.

A strong converging lens makes a good magnifying glass. It needs to be held fairly close to the object to give an image (closer than the focal length), but if it is too close the magnification will be small.

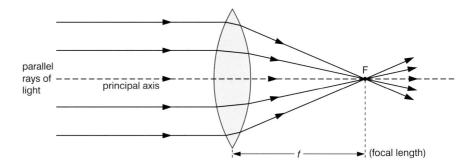

parallel rays of light — principal axis — F — *f* — (focal length)

If the object is further from the lens than the focal length, the converging lens forms a real image. For example the cinema projector makes a real image on the screen of the film inside the projector, and the camera makes a real image of the object being photographed on the film or the digital sensor inside. In the first case the image is bigger than the object, in the second case it is smaller.

TO FIND THE POSITION OF AN IMAGE

To find the position of the image of an object formed by a lens, we can draw a ray **diagram**. There are three standard rays that we can use to do this. A standard ray is one whose complete path we know. In ray diagrams, any two of the standard rays are needed to find the position and size of the image, although it is a wise precaution to draw the third ray, as a check on the accuracy with which the first two were drawn.

For convenience, ray diagrams are usually drawn with the bottom of the object on the principal axis. This means that the bottom of the image is also on the principal axis, and we need only locate the top of the image.

The standard rays

1 A ray from the top of the object, straight through the centre of the lens.

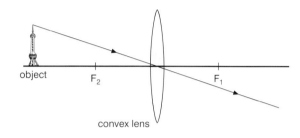

2 A ray from the top of the object, parallel to the principal axis until it reaches the lens, and then down through the principal focus F_1 on the far side of the lens.

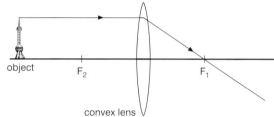

3 A ray from the top of the object through the principal focus F_2 on the near side of the lens, down to the lens and then parallel to the axis.

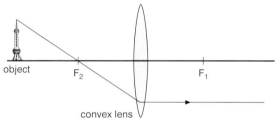

By using a suitable combination of these three rays, images can be located. Three examples are given below.

Object a long way from the lens (more that twice the focal length)

Here the image is real, inverted, smaller than the object and closer to the lens than the distance of the object from the lens. A real image is one through which the rays actually pass, and which could be picked up on a suitable screen.

Examples of the formation of this type of image are in the eye and in the camera.

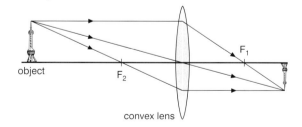

Object closer to the lens (between 1x and 2x the focal length)

Here the image is real, inverted, larger than the object and further from the lens than the distance of the object from the lens.

Examples of the formation of this type of image are in the film projector and in the photographic enlarger.

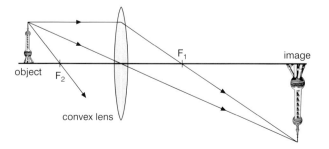

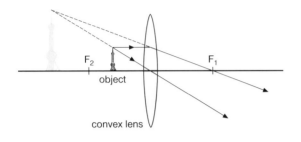

convex lens

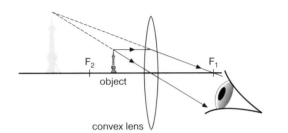

convex lens

Object closer to the lens than the focal length

Here the image is virtual, upright, larger than the object and further away from the lens than the distance of the object from the lens. A **virtual image** is one through which the rays do not actually pass, but from which they appear to come. Such an image cannot be picked up on a screen.

An example where this type of image is formed is in the magnifying glass. The diagram below shows how the eye sees the image through a magnifying glass.

Note that to see the image, you have to look through the lens.

Note also that, as in the case of the mirror on page 91, the image in all these cases is never on the lens itself. This is a mistake which is often made by students in examinations.

WORKED EXAMPLE

Find the position of the image in the following diagram.

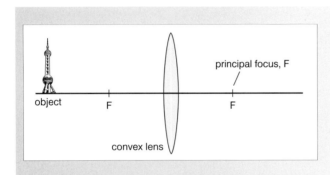

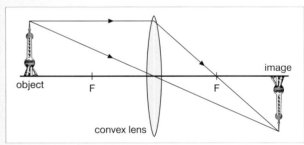

A* EXTRA

• If the diagram is drawn to scale, the size of the image can also be found.

The camera forms a small inverted image of the object in front of it on the digital sensor or the film at the back.

Note that these special rays are 'construction lines'. In particular if the lens has a smaller diameter, the upper ray could miss the lens completely. This does not matter, simply pretend that the lens has a large enough diameter, and work out where the image is. That is where all of the real rays will go. The image will not move just because the real lens has a smaller diameter.

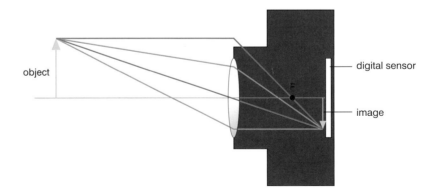

As you can see, we use the two rays shown in blue to find where the image would be. It is clear that one of the blue rays cannot really be followed by light because the lens is too small. The grey lines are some of the real light rays that go through the lens, but these rays cannot be used to find the position of the image.

Dispersion of light

If white light is passed through a triangular prism (a glass or plastic block), it is split into a spectrum of different colours. This is called dispersion. The refractive index of the glass or plastic is slightly different for each colour of light, so each colour is refracted by a different amount. The light rays are bent or refracted twice: once as they enter the prism, and again in the same direction as they leave it. The dispersion increases each time, which is why the colours are separated so much. Red is always bent least, and violet is bent most.

Electromagnetic spectrum

The **electromagnetic spectrum** is a 'family' of waves. Electromagnetic waves all travel at the same speed in a vacuum, i.e. the speed of light, 300 000 000 m/s. This can be written more conveniently as 3×10^8 m/s. This high speed explains why you can have a phone call between China and New Zealand with only an extra delay of 0.1 s before you hear the reply from the person at the other end. It takes the infrared signal this long to travel there and back through an optical fibre.

However, for astronomical distances the delays quickly become longer. Even when Mars is at its nearest to Earth, it takes 10 minutes to send a message to a robot on the surface and receive a reply. Getting a reply from the nearest star would take $8\frac{1}{2}$ years.

Note that all electromagnetic waves can travel through a vacuum, which is why we can see the light and feel the heat coming from the Sun. Other waves, such as sound waves, cannot travel through a vacuum.

White light is a mixture of different colours and can be split by a prism into the **visible spectrum**. All the different colours of light travel at the same speed in a vacuum but they have different frequencies and wavelengths. Red light has a wavelength that is about twice as long as violet light. When light waves of different frequencies enter glass or perspex they all slow down, but by different amounts. The different colours are therefore refracted through different angles. Violet is refracted the most, red the least.

The visible spectrum is only a small part of the full electromagnetic spectrum.

Light of one wavelength, that is to say of just one colour, is known as **monochromatic** light.

A prism splits white light into the colourful spectrum of visible light.

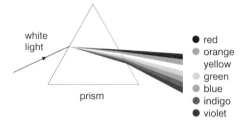

Type of wave	gamma rays	X-rays	ultraviolet	visible	infrared	microwaves	TV and radio waves
frequency	high						low
wavelength	low						high
use	killing cancer cells	to look at bones	suntan beds	photography	TV remote controls	cooking	transmission of TV and radio

The complete electromagnetic spectrum.

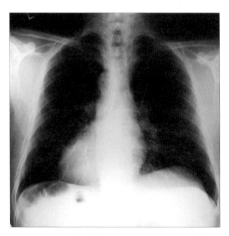

An X-ray may save your life, but it is not free of risk. An X-ray is used when a doctor is sure that the benefits are vastly greater than the risks, which are very small but are not zero.

A TV remote control uses pulses of IR light to send signals to the TV.

THE WAVES YOU CANNOT SEE

Gamma rays are produced by radioactive nuclei. They transfer more energy than X-rays and can cause cancer or mutation in body cells. Gamma rays are frequently used in radiotherapy to kill cancer cells. The success of this treatment is greatly improved by the fact that cancer cells are easier to kill than ordinary cells.

Radioactive substances that emit gamma rays are used as tracers. For example, if scientists want to know where in a plant the phosphorus goes, they can feed it a radioactive isotope of phosphorus and then measure the activity that comes out of different parts of the plant.

X-rays are produced when high-energy electrons are fired at a metal target. Bones absorb more X-rays than other body tissue. If a person is placed between the X-ray source and a photographic plate, the bones appear to be white on the developed photographic plate compared with the rest of the body. X-rays have very high energy and can damage or destroy body cells. They may also cause cancer. X-rays are also used to treat cancer.

Ultraviolet radiation (UV) is the component of the Sun's rays that gives you a suntan. UV is also created in fluorescent light tubes by exciting the atoms in a mercury vapour. The UV radiation is then absorbed by the coating on the inside of the fluorescent tube and re-emitted as visible light. Fluorescent tubes are more efficient than light bulbs because they do not depend on heating and so more energy is available to produce light. Ultraviolet can also damage the surface cells of the body, which can lead to skin cancer, and can damage the eyes, leading to blindness.

All objects give out **infrared radiation** (IR). The hotter the object is, the more radiation it gives out. Thermograms are photographs taken to show the infrared radiation given out from objects. Infrared radiation grills and cooks our food in an ordinary oven and is used in remote controls to operate televisions and videos. Infrared can burn skin and other body tissue.

Microwaves are high-frequency radio waves. They are used in radar to find the position of aeroplanes and ships. Metal objects reflect the microwaves back to the transmitter, enabling the distance between the object and the transmitter to be calculated. Microwaves are also used for cooking. Water particles in the food absorb the energy carried by microwaves. They vibrate more, and this vibration is spread to all of the atoms in the food. The increased vibration can be measured with a thermometer as an increase in temperature. Microwaves penetrate several centimetres into the food and so speed up the cooking process. Because of this, microwaves can heat body tissue internally.

Radio waves have the longest wavelengths and lowest frequencies. **UHF** (ultra-high frequency) waves are used to transmit television programmes to homes. **VHF** (very high frequency) waves are used to transmit local radio programmes. **Medium** and **long** radio waves are used to transmit over longer distances because their wavelengths allow them to diffract around obstacles such as buildings and hills. Communication satellites above the Earth receive signals carried by high-frequency (**short-wave**) radio waves. These signals are amplified and re-transmitted to other parts of the world.

REVIEW QUESTIONS

Q1 This is a list of types of wave:
gamma infrared light microwaves radio ultraviolet X-rays
Choose from the list the type of wave that best fits each of these descriptions.
  a Stimulates the sensitive cells at the back of the eye.
  b Necessary for a suntan.
  c Used for rapid cooking in an oven.
  d Used to take a photograph of the bones in a broken arm.
  e Emitted by a video remote control unit.

Q2 **a** Rays of light can be reflected and refracted. State one difference between reflection and refraction.
  b The diagram shows a glass block and two rays of light.
    i Complete the paths of the two rays as they pass into and then out of the glass block.
    ii What name is given to the angle marked *a*?
    iii What name is given to the line marked XY?

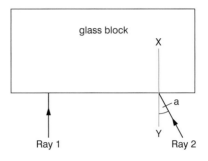

Q3 The diagram shows light entering a prism. Total internal reflection takes place at the inner surfaces of the prism.
  a Complete the path of the ray.
  b Suggest one use for a prism like this.
  c Complete the table about total internal reflection. Use T for total internal reflection or R for refraction.

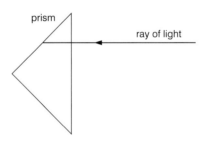

angle of incidence/°	total internal reflection (T) or refraction (R)
36	
42 (critical angle)	
46	

Q4 A student traces the path of a red light beam through a rectangular block of plastic, and finds that the angle of incidence is 50°, and the angle of refraction is 21.7°.
  a What is the refractive index of the block?
  b What can you say about the angle of refraction for a blue light beam with an angle of incidence of 50°?
  c What will be the critical angle for this material?

Examination questions are on page 104.

More questions on the CD ROM

3 Sound

Sound travels much more slowly than light in the air. We can use this to measure how quickly the sound does travel.

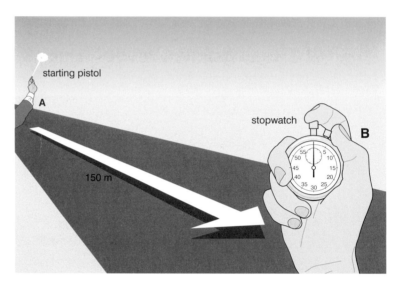

Person A makes a loud sound and produces a visual signal at the same time – this could be by firing a starting pistol or by banging large cymbals together. Person B starts a stopwatch when they *see* the sound being made and stops the stopwatch when they *hear* the sound. They can work out the speed of sound using this formula:

$$\text{speed of sound} = \frac{\text{distance between the two people/m}}{\text{time measured/s}}$$

Properties of sound waves

Sound waves travel at about 340 m/s in the air – much slower than the speed of light. This explains why you almost always see the flash of lightning before hearing the crash of the thunder.

Sound is caused by vibrations, of the front of a violin or a cello, or of the column of air inside a trumpet. In the case of a loudspeaker you can see that the cone of the loudspeaker moves in and out and changes the pressure in the air in front of it. The sound travels as **longitudinal waves**. The compressions and rarefactions of sound waves result in small differences in air pressure.

Sound waves travel faster through liquids than through air. Sound travels fastest through solids. This is because particles are linked most strongly in solids. Note, however, that sound must have a medium through which to travel. Unlike electromagnetic waves, sound will not travel through a vacuum.

High-**pitch** sounds have a high frequency. Examples of high-pitch sounds include bird-song, and all the sounds that you hear from someone else's personal player when they have set the volume too high. Low-pitch sounds have a low frequency. Examples of low-pitch sounds include the horn of a large ship and the bass guitar.

The human ear can detect sounds with pitches ranging from 20 Hz to 20 000 Hz. Sound with frequencies above this range is known as **ultrasound**. Ultrasound is used by bats for navigation and by doctors for looking at unborn babies.

The ear is far more easily damaged than most people realise, and care needs to be taken both with the volume of sound and the length of time that the ear is exposed to it. The damage is cumulative, and so you don't notice it at first. Many older rock stars have serious hearing problems, and younger ones often wear ear plugs.

Loud sounds have high amplitude whereas quiet sounds have low amplitude. The unit in which we measure the loudness of sounds is the decibel (dB). Decibels are used to measure various electrical quantities as well, but when they are used for sound, 0 dB is defined to be the quietest sound that can be heard. This then makes a quiet room at night about 40 dB; a noisy classroom is 60 dB; the sound 1 m from a vacuum cleaner is 80 dB; a loud disco could be 100 dB, which would be illegal in many countries; sounds of 120 dB are extremely painful; and windows break at 160 dB.

The ears start to be damaged at 85 dB, a level that some personal players can reach without difficulty.

Sound waves can be displayed on an oscilloscope by using a microphone. This produces a voltage-time graph on the oscilloscope.

This orchestra is creating a single longitudinal wave of very complicated shape. In ways that we barely understand, our brains can pick out the sounds of all the individual instruments that are playing together.

A* EXTRA

- The speed of sound in water is about 1500 m/s, and in hard metals and wood it is about 5000 m/s.

vibrating loudspeaker

microphone

oscilloscope

amplitude

Typical results are shown on the graphs below.
(a) A loud sound of low frequency
(b) A loud sound of high frequency
(c) A quiet sound of low frequency
(d) A quiet sound of high frequency.

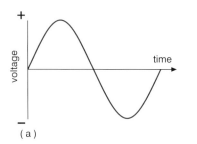

(a)

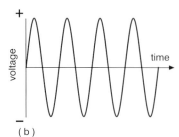

(b)

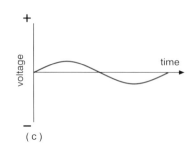

(c)

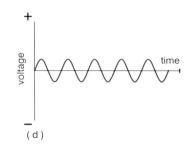

(d)

WORKED EXAMPLE

Using this displacement–time graph for a sound wave, calculate:
a the amplitude of the sound
b the frequency of the sound.

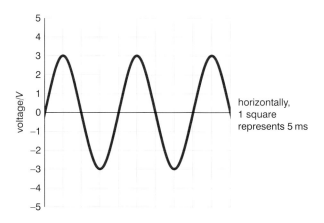

horizontally, 1 square represents 5 ms

a The amplitude is the maximum displacement from the mean position, so can be read straight from the graph.

Amplitude = 3 mm.

b From the graph, work out the time taken for one cycle:

One cycle covers 4 squares and each square represents 5 ms, so the time for one cycle is 20 ms.

Convert to seconds: 1 ms = 0.001 seconds, so 20 ms = 0.020 s

The frequency is the number of cycles per second = $\frac{1}{0.020}$ = 50 Hz.

ECHOES

Hard surfaces reflect sound waves. An echo is a sound that has been reflected before you hear it. For an echo to be clearly heard, the obstacle needs to be large compared with the wavelength of the sound. So, you will hear an echo if you make a loud noise when you are several hundred metres from a brick wall or a cliff, for example. You will not hear an echo if you are several hundred metres from a pole stuck in the ground. There will still be an echo, even if you are much closer to the wall, but because sound travels very quickly, the echo will return in such a short time that you will probably not be able to distinguish it from the sound that caused it.

Measuring the speed of sound by an echo method

The following worked example illustrates how echoes may be used to measure the speed of sound.

WORKED EXAMPLE

Two students stand side by side at a distance of 480 m from the school wall. Student A has two flat pieces of wood, which make a loud sound when clapped together. Student B has a stopwatch.

As student A claps the boards together, student B starts the stopwatch. When student B hears the echo, he stops the stopwatch. The time recorded on the stopwatch is 2.9 s.

Calculate the speed of sound.

Write down the formula: $\qquad$ speed of sound $= \dfrac{\text{distance travelled}}{\text{time taken}}$

Work out the distance: $\qquad$ distance to wall and back $= 2 \times 480$
$= 960$ m

Record the time the sound took $\qquad$ time $= 2.9$ s
to travel there and back:

Substitute in the formula: $\qquad$ speed of sound $= \dfrac{960}{2.9}$
$= 331$ m/s

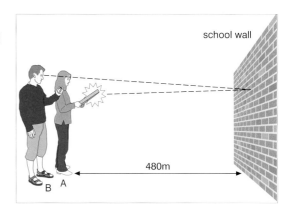

See if you can think of some things which might be done to improve the accuracy of this experiment.

REVIEW QUESTIONS

Q1 **a** i What causes a sound?
　　ii Explain how sound travels through the air.
　b Astronauts in space cannot talk directly to each other. They have to speak to each other by radio. Explain why this is so.
　c If a marching band is approaching you, explain why you can hear the bass drum long before you can hear the piccolo playing the highest notes.

Q2 Ayesha and Salma are doing an experiment to measure the speed of sound. They stand 150 m apart. Ayesha starts the stopwatch when she sees Salma make a sound and she stops it when she hears the sound herself. She measures the time as 0.44 seconds. Calculate the speed of sound in air from this data.

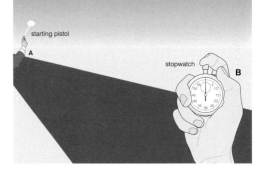

Q3 The speed of sound is approximately 340 m/s.
　a Calculate the wavelength of middle C, which has a frequency of 256 Hz.
　b A student hears two echoes when she claps her hands. One echo is 0.5 s after the clap, and one echo is 1.0 s after the clap. She decides that the two echoes are from two buildings in front of her. How far apart are the buildings?

More questions on the CD ROM

Examination questions are on page 104.

EXAMINATION QUESTIONS

Q1 Fig. 1.1 shows the diffraction of waves by a narrow gap.
P is a wavefront that has passed through the gap.

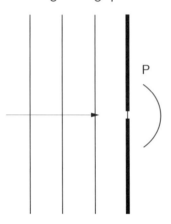

Fig. 1.1

    a On Fig. 1.1, draw three more wavefronts to the right of the gap.   [3]

    b The waves travel towards the gap at a speed of 3×10^8 m/s and have a frequency
       of 5×10^{14} Hz. Calculate the wavelength of these waves.
       wavelength = _____ [3]

Q2 a Fig. 2.1 shows the air pressure variation along a sound wave.

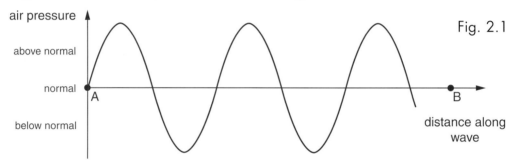

Fig. 2.1

      i On AB in Fig. 2.1, mark one point of compression with a dot and the letter C and
        the next point of rarefaction with a dot and the letter R.
      ii In terms of the wavelength, what is the distance along the wave between a
        compression and the next rarefaction?
      _____ [3]

    b A sound wave travels through air at a speed of 340 m/s. Calculate the frequency
      of a sound wave of wavelength 1.3 m.
      frequency = _____ [2]

Q3 Fig. 3.1 shows the cone of a loudspeaker that is producing sound waves in air.
At any given moment, a series of compressions and rarefactions exists along the line XY.

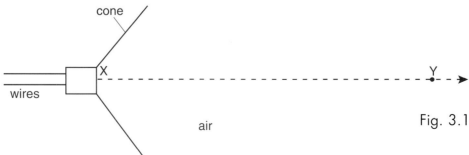

Fig. 3.1

a   On Fig. 3.1, use the letter C to mark **three** compressions and the letter R to mark
    three rarefactions along XY.   [1]

b   Explain what is meant by
    i   a *compression*,
    _____
    _____

    ii a *rarefaction*.
    _____
    _____ [2]

c   A sound wave is a longitudinal wave. With reference to the sound wave travelling
    along XY in Fig. 3.1, explain what is meant by a *longitudinal* wave.
    _____ [2]

d   There is a large vertical wall 50 m in front of the loudspeaker. The wall reflects the
    sound waves.
    The speed of sound in air is 340 m/s.
    Calculate the time taken for the sound waves to travel from X to the wall and to
    return to X.
    time = _____   [2]

Q4 Fig. 4.1 shows wavefronts of light crossing the edge of a glass block from air into glass.

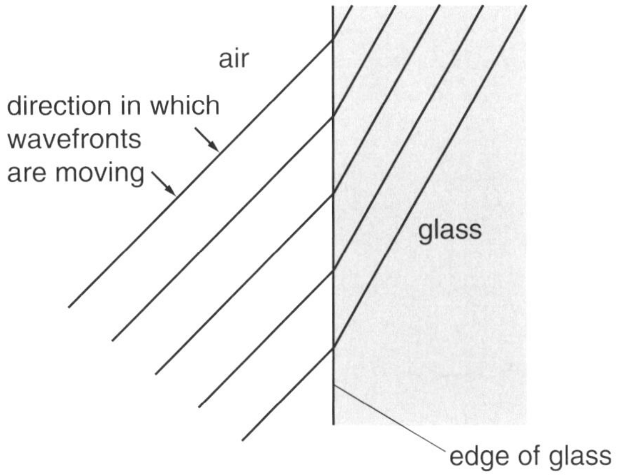

Fig. 4.1

a   On Fig. 4.1
    i   draw in an incident ray, a normal and a refracted ray that meet at the same
         point on the edge of the glass block,
    ii   label the angle of incidence and the angle of refraction,
    iii measure the two angles and record their values.
         angle of incidence = _____
         angle of refraction = _____   [4]

b   Calculate the refractive index of the glass.
    refractive index = _____   [3]

The Arctic Tern travels over 14,000 km in a year from the Arctic to the Antarctic and back again

Animal magnetism

How do migrating animals find their way? Arctic terns travel from the North Pole to the South Pole and back again each year; whales migrate from Hawaii to the northern Pacific coast and back. So why don't they get lost? A number of studies suggest that many animals have a magnetic sense that allows them to 'tune in' to the Earth's magnetic field.

The Earth behaves as if it has a giant bar magnet at its centre. Of course, there isn't really a rectangular block at the centre of the Earth, but the magnetic field does provide evidence that there is iron at the core. This magnetic core produces a magnetic field that extends through the surface and all around us. When we use a compass we are detecting these magnetic field lines. A material called biomagnetite has been discovered in the brains of a number of animals and it is thought that this allows the animals to sense the Earth's magnetic field and so find their way.

Animals use other techniques as well, such as sonar and sighting landmarks along the way, but it seems that some animals can use a compass, just like us. It's just that they carry their compass with them inside their bodies.

ELECTRICITY AND MAGNETISM

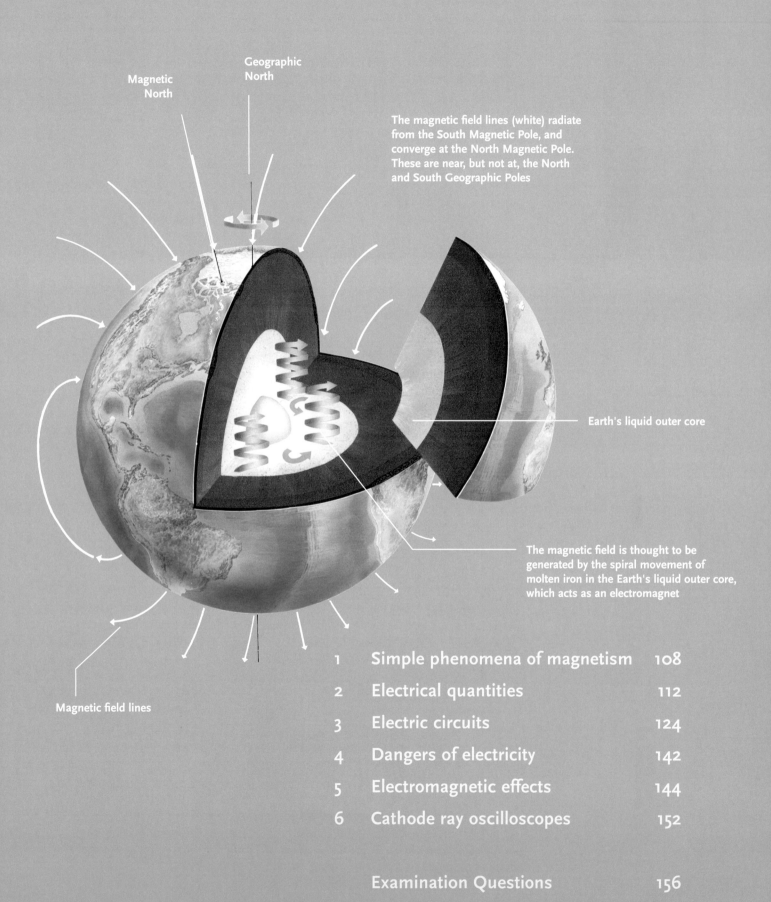

Magnetic North

Geographic North

The magnetic field lines (white) radiate from the South Magnetic Pole, and converge at the North Magnetic Pole. These are near, but not at, the North and South Geographic Poles

Earth's liquid outer core

The magnetic field is thought to be generated by the spiral movement of molten iron in the Earth's liquid outer core, which acts as an electromagnet

Magnetic field lines

1 SIMPLE PHENOMENA OF MAGNETISM

Videos & questions on the CD ROM

There are several elements that are magnetic, the most important of which are iron, cobalt and nickel. Scientists have developed alloys and ceramics made from complicated combinations of elements to get the exact properties that they want. Some of these materials are **magnetically hard** (such as steel, which is an alloy of iron, and other elements such as carbon or tungsten). This means that they stay magnetic once they have been magnetised.

When we refer to a 'magnet', we mean **a permanent magnet** that is made of magnetically hard materials.

Other materials are **magnetically soft** (such as pure iron), which means that they do not stay magnetic – this is particularly useful in some electromagnetic devices such as the electromagnet and the relay.

Alloys are made by melting different metallic elements (iron, aluminium, copper, tungsten, etc.) together. The resulting metal is known as a **ferrous** metal if it contains a lot of iron, and as a **non–ferrous** metal if it does not. Nickel and brass (copper + tin) are examples of non-ferrous metals.

In the past all magnetic materials were ferrous, but this is no longer true, and the strongest magnets may not contain any iron at all, for example, samarium-cobalt (SmCo), often used in headphones.

When we refer to 'magnetically hard' and 'magnetically soft' materials we are not referring to their physical hardness. You may have seen rubberised magnetic strips used on notice boards. These strips are permanent magnets, but are physically soft.

If a permanent magnet is suspended and allowed to swing, it will line up approximately north–south. Because of this, the two ends of a magnet (which are the most strongly magnetic parts) are called the north pole and the south pole, often labelled N and S. (Strictly, they are called the north-seeking pole and the south-seeking pole.)

If two north poles from different magnets are brought together, there will be a **repulsion** between them. The same happens if two south poles are brought together. However, if a north pole and a south pole are brought together, there will be an **attraction**.

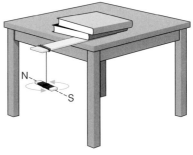

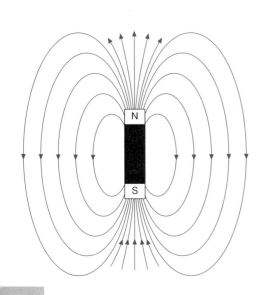

Magnets have a **magnetic field** around them – a region of space where their magnetism affects other objects. We describe the magnetic field using **magnetic field lines**. These lines show the path that the north pole of a magnet would take: heading away from a north pole and ending up at a south pole. The more concentrated the field lines are, the stronger the magnetic effect.

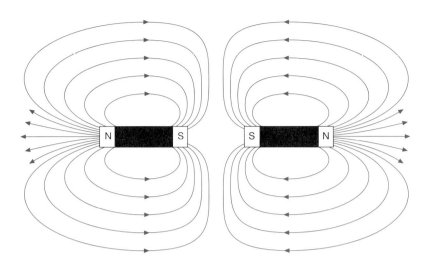

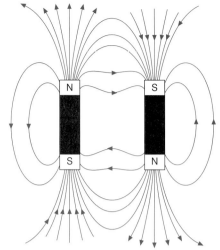

The magnetic field around a pair of bar magnets depends on the orientation of the magnets.

AN EXPERIMENT TO SHOW THE FIELD LINES AROUND A BAR MAGNET

To show the field lines, place a bar magnet under a thin sheet of plastic, and sprinkle iron filings on to the top of the plastic. The iron filings will arrange themselves into strings of filings along the field lines.

It is also possible to follow the path of the field lines by placing a small compass (known as a plotting compass) on the plastic in place of the iron filings. If you move the compass in the direction that its north pole is pointing, then it will follow a field line.

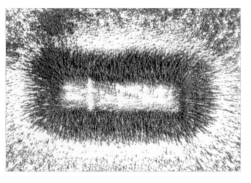

The iron filings show the field lines around a bar magnet.

If a soft magnetic material is brought near to a magnet it will be attracted. It has had magnetism **induced** in it, it has become **magnetised**. When the magnet is taken away, the material loses its magnetism again. Note that the magnet will continue to attract the soft magnetic material even if the material is turned round. This is the opposite behaviour to two magnets, as two magnets will repel each other in certain orientations. This simple method enables you to work out whether you are holding two magnets or one magnet and one piece of soft magnetic material.

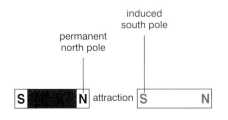

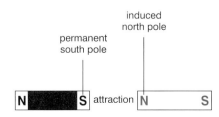

The pole of a permanent magnet always induces the opposite pole in an unmagnetised piece of magnetic material. So an induced magnet is always attracted to a permanent magnet.

Horseshoe magnet.

TYPES OF MAGNET

Bar magnets, as shown on page 109, are used in compasses.

Horseshoe magnets bring the N and S poles close together, and this gives a strong magnetic field between the poles. They are used to lift heavy weights. With older magnetic materials it is necessary to place a 'keeper' across the poles when the magnet is stored to prevent the loss of magnetism.

The magnet used in a loudspeaker is very similar to the horseshoe magnet. The central pillar is made of one pole, and on the other side of a small gap, the opposite pole surrounds it. The coil of the loudspeaker moves in this gap.

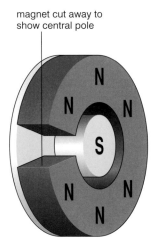

Magnet in a loudspeaker.

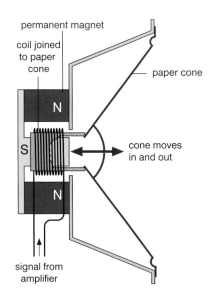

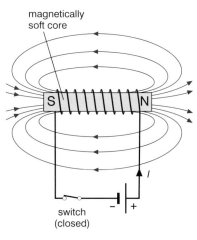

Electromagnet.

Electromagnets are made out of a coil of wire (often called a solenoid). When an electric current is passed through the coil, a magnet is formed with the N pole at one end of the coil and the S pole at the other end. If the coil is wrapped around a magnetically soft core, then when the coil is magnetised, it magnetises the core as well, and a very much stronger magnetic field is made. When the current is switched off, the coil loses its magnetism, so the core does as well.

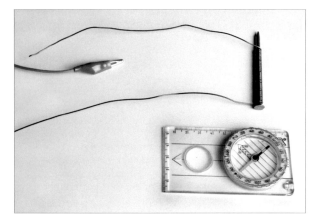

No current is flowing through the coil, and the compass points to the north.

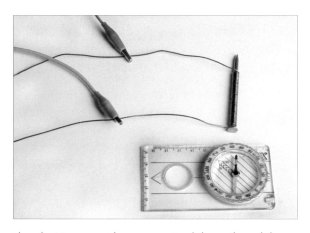

The electric current has magnetised the coil, and the coil has magnetised the soft iron core. The compass has lined itself up to the magnetic field.

METHODS OF MAGNETISATION

Permanent magnets are usually magnetised by putting them into a coil of wire and passing a large direct current of electricity through the coil for a moment. Very high currents are used, and special equipment must be used to make the method safe.

With some modern alloys, the magnetic field is applied to the molten alloy at very high temperature, and then the alloy is allowed to solidify while remaining in the magnetic field.

Steel can be magnetised to a certain degree by placing the bar N→S and hammering it. Ships become magnetised in this way during manufacture, as the hammering allows the Earth's field to magnetise them slightly.

METHODS OF DEMAGNETISATION

The only method of demagnetisation that is guaranteed to work is to heat the magnet. All magnets have a temperature (called the Curie temperature) at which they lose their magnetism. This temperature ranges from less than 100 °C to over 500 °C. When the magnet is cooled down, some of the magnetism may return.

Many magnets can be demagnetised by placing them in a coil of wire connected to an alternating electric current (a.c.) source. The current is switched on and while the a.c. current is flowing, the object is slowly taken out of the coil. In this way the object is magnetised in the opposite direction each time the current reverses, but as it is removed from the coil the amount of magnetisation is reduced each time.

Hammering or even dropping a powerful magnet may cause the loss of some of its magnetisation.

REVIEW QUESTIONS

Q1 What is the difference between a magnetically hard material and a magnetically soft material? Give an example of each.

Q2 Ranjit has a piece of metal that he thinks is a magnet. He holds it near another magnet and it is attracted. Ranjit says this proves his metal is a magnet. Explain why Ranjit is wrong.

Q3 Sketch the magnetic field pattern for a single bar magnet. How would the diagram change if the magnet were made stronger?

More questions on the CD ROM

Examination questions are on page 156.

2 ELECTRICAL QUANTITIES

Videos & questions on the CD ROM

Electric charge

All atoms are made up of three kinds of particles, called **electrons**, **protons** and **neutrons**. Electrons are the tiniest of these, and have a negative charge. Protons and neutrons have about the same mass, but protons are positively charged, while neutrons have no charge.

In most objects there are as many electrons as protons. So normally an object has no overall charge, because the positive charge on all the protons is matched by the negative charge on the electrons. If there are more electrons than protons the object carries an overall negative charge. If there are fewer electrons than protons, the object carries an overall positive charge.

Every proton and electron produces an electric field. So around any object in which the charges are not balanced, there is an electric field. When a charged particle moves into the field, it feels a force towards or away from the other particle (see below). The strength of the force depends on:

• how close the particles are: the closer they are, the larger the force

• how much electrical charge they carry: the more charge, the larger the force.

Field lines show the shape of an electric field.

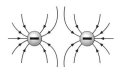 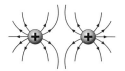 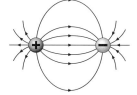

Like charges repel each other.　　　　Unlike charges attract each other.

Because the static charge on each hair is similar, the hairs repel each other and stick up in all directions.

When an unbalanced charge collects on the surface of an object, the charge is called **static charge**. ('Static' means 'not moving'.) When electrons move, or flow, from one place to another, they produce an electric current.

As you can see, these forces, which are called **electrostatic forces**, look rather similar to magnetic forces. They are however completely different. An electric field does not affect a magnet in any special way, and a magnetic field does not affect an electric charge (so long as it is *not moving*). You can even have a space that contains both types of field in different directions at the same time.

Electric fields and electrostatics feature in our lives. They can be useful: electrostatic scrubbers remove the dust from the smoke of coal power stations, and photocopiers use electrostatics to move the ink powder to the right place on the paper. But they can also be harmful. You may have noticed that you can get a nasty spark from your finger if you touch a metal object after rubbing your feet on a nylon carpet. For this reason, aircraft are connected to the ground by a special wire before refuelling;

any electrostatic charges can thus flow away safely to earth and not cause a spark. The charges might be built up by friction between the fuel and the fuel pipe.

When you charge an object you are giving or taking away negatively charged electrons, so that the charge on the object overall is unbalanced. For example, when you rub a glass or acetate rod with a cloth, electrons from the rod get rubbed onto the cloth (see diagram below). So the cloth becomes negatively charged overall, and the rod is left with an overall positive charge. When you rub a polythene rod with a cloth, electrons from the cloth get transferred to the rod, so the polythene carries a negative charge overall, and the cloth carries a positive charge.

If you suspended charged polythene and acetate rods so they could move freely, and brought the two close together, they would attract each other, since unlike charges attract. Both the polythene and the acetate rod would attract small bits of paper or dust, because they give each bit an opposite charge by induction (see diagram below).

Materials like glass, acetate and polythene can only become charged because they are insulators. Electrons do not move easily through insulating materials, so when extra electrons are added, they stay on the surface instead of flowing away, and the surface stays negatively charged. Similarly, if electrons are removed, electrons from other parts of the material do not flow in to replace them, so the surface stays positively charged. A material through which electrons flow easily is called a conductor. Conductors, such as metals, cannot be charged by rubbing.

You can show how much charge is on an object, and whether it is positive or negative, using a gold leaf electroscope (see below).

A* EXTRA

- The amount of charge on an object is measured in coulombs (C). A charge of 1 C is the charge on 6.2×10^{18} electrons, so an object with a charge of +1.0 C has these many too *few* electrons. This is an enormous charge, and objects would explode long before they could be given so much. Typical electrostatic charges are less than 1 μC (1.0×10^{-6} C).

Charging an electroscope by induction.

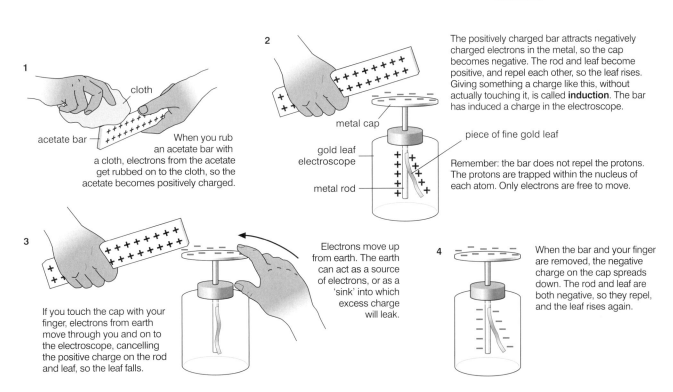

1 cloth

acetate bar

When you rub an acetate bar with a cloth, electrons from the acetate get rubbed on to the cloth, so the acetate becomes positively charged.

2 metal cap

gold leaf electroscope

metal rod

piece of fine gold leaf

The positively charged bar attracts negatively charged electrons in the metal, so the cap becomes negative. The rod and leaf become positive, and repel each other, so the leaf rises. Giving something a charge like this, without actually touching it, is called **induction**. The bar has induced a charge in the electroscope.

Remember: the bar does not repel the protons. The protons are trapped within the nucleus of each atom. Only electrons are free to move.

3 If you touch the cap with your finger, electrons from earth move through you and on to the electroscope, cancelling the positive charge on the rod and leaf, so the leaf falls.

Electrons move up from earth. The earth can act as a source of electrons, or as a 'sink' into which excess charge will leak.

4 When the bar and your finger are removed, the negative charge on the cap spreads down. The rod and leaf are both negative, so they repel, and the leaf rises again.

Testing charge

1

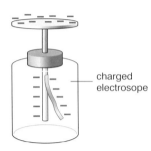

charged electrosope

You can use an electroscope to test if an object is positively or negatively charged.

2

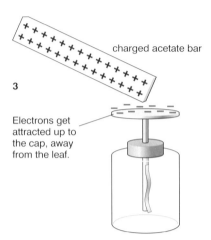

charged polythene bar

If you bring an object with the same charge as the electroscope close to the cap, the leaf rises further. The more charge on the object, the further the leaf rises.

3

charged acetate bar

Electrons get attracted up to the cap, away from the leaf.

If you bring an object with the opposite charge to the electroscope close to the cap, the leaf falls. The more charge on the object, the further the leaf falls.

Lightning is a spectacular example of electrostatics in action. We believe that the electrical charge is generated by induction when ice particles in the clouds collide. One bolt of lightning is about 5 C of electrical charge. Lightning conductors on buildings usually prevent lightning strikes by releasing the opposite charge into the cloud above to neutralise the charge that is building up.

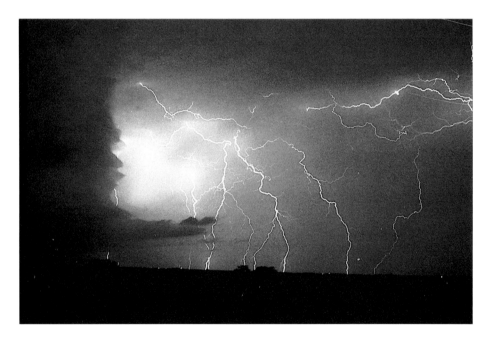

Current

All materials contain electrons, but in many materials they are all 'locked' into the material's atoms and cannot move about. These materials cannot carry an electric current, and are called electrical insulators. Materials in which there are large numbers of electrons that are free to move around from atom to atom are called conductors.

When there is no current in a conductor, the free electrons move randomly between atoms, with no overall movement. When you connect it in an electrical circuit with a power source like a battery, there is a current in the conductor. Now the electrons drift in one direction, while still moving in a random way as well. The drift speed is very slow, often only a few millimetres each second. A current can only flow in a conductor if it is connected in a complete circuit. If the circuit is broken, the current stops.

The size of an electric current depends on the number of electrons that are moving and how fast they are moving. But instead of measuring the actual number of electrons we use the total charge carried by the electrons round the circuit each second.

Electric current is measured in **amperes**, or **amps** (A).

If there is a current of 1 A in a wire, then one coulomb of charge is passing any point on the circuit each second. (1 A = 1 C/s.)

Yes, this coulomb is the same one that we referred to earlier on. The coulomb of charge is vastly safer when it is made of electrons flowing along a conducting wire because it is then closely surrounded by positive charges, and there are no large electric fields.

You use an ammeter to measure current in an electrical circuit. If the current is very small, you might use a milliammeter, which measures current in milliamps (1 mA = 0.001 A). Even smaller currents are measured with a microammeter.

If you want to measure the current in a particular component, such as a lamp or motor, the ammeter must be connected **in series** with the component. In a series circuit, the current is the same no matter where the ammeter is put. This is not the case with a parallel circuit.

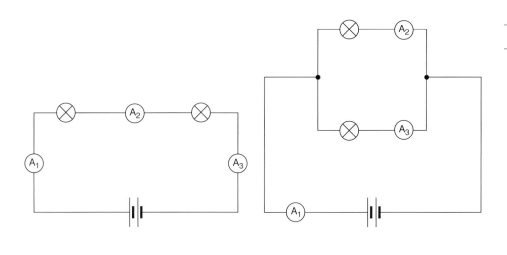

In this series circuit, the current will be the same throughout the circuit so $A_1 = A_2 = A_3$.

The current splits between the two branches of the parallel circuit so $A_1 = A_2 + A_3$.

The electric current is the amount of charge flowing every second – the number of coulombs per second:

$I = \dfrac{Q}{t}$

I = current in amperes (A)

Q = charge in coulombs (C)

t = time in seconds (s)

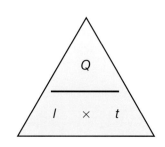

Electro-motive force

The battery in an electrical circuit can be thought of as pushing electrical charge round the circuit to make a current. It also transfers energy to the electrical charge. The **electro–motive force** (e.m.f.) of the battery, measured in volts, measures how much 'push' it can provide and how much energy it can transfer to the charge.

Scientists now know that electric current is really a **flow of electrons** around the circuit from negative to positive. Unfortunately, early scientists guessed the direction of flow incorrectly. Consequently all diagrams were drawn showing the current flowing from positive to negative. This way of showing the current has not been changed and so the **conventional current** that everyone uses gives the direction that positive charges would flow.

Conventional current is drawn in the opposite direction to electron flow.

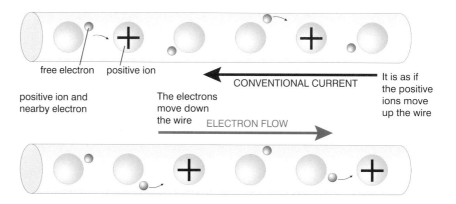

free electron   positive ion

positive ion and
nearby electron

The electrons
move down
the wire   ELECTRON FLOW

CONVENTIONAL CURRENT

It is as if
the positive
ions move
up the wire

Potential difference

The electrons moving round a circuit have some **potential energy**. As electrons move around a circuit, they transfer energy to the various components in the circuit. For example, when the electrons move through a lamp they transfer some of their energy to the lamp.

The amount of energy that a unit of charge (a coulomb) transfers between one point and another (the number of joules per coulomb) is called the **potential difference** (p.d.). Potential difference is measured in **volts** and so it is often referred to as **voltage**.

If the potential difference across a lamp, say, is 1 volt, then each coulomb of charge that passes through the lamp will transfer 1 joule of energy to the lamp.

A* EXTRA

• The potential difference is measured between two points in a circuit. It is like an electrical pressure difference and measures the energy transferred per unit of charge flowing.

Potential difference (p.d.) is the difference in energy of a coulomb of charge between two parts of a circuit.

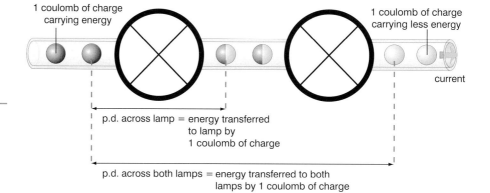

1 coulomb of charge
carrying energy

1 coulomb of charge
carrying less energy

current

p.d. across lamp = energy transferred
to lamp by
1 coulomb of charge

p.d. across both lamps = energy transferred to both
lamps by 1 coulomb of charge

MEASURING ELECTRICITY

Potential difference is measured using a **voltmeter**. If you want to measure the p.d. across a component then the voltmeter must be connected across that component. Testing with a voltmeter does not interfere with the circuit.

A voltmeter can be used to show how the potential difference varies in different parts of a circuit. In a series circuit you find different values of the voltage depending on where you attach the voltmeter. You can assume that energy is only transferred when the current passes through electrical components such as lamps and motors – the energy transfer as the current flows through copper connecting wire is very small. It is only possible therefore to measure a p.d. or voltage across a component.

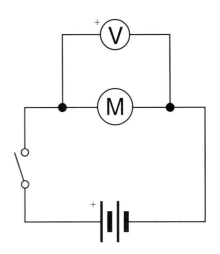

The voltmeter can be added after the circuit has been made.

p.d. = 6 V
    6 joules transferred from each coulomb

power = 2 coulombs per second, of 6 joules each
      12 joules per second

p.d. = 3 V
    = 3 J/C

power = 2 C/s × 3 J/C
      = 6 joules per second

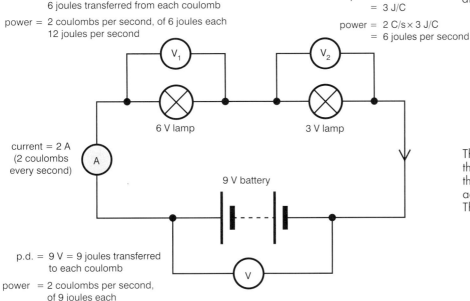

current = 2 A
(2 coulombs every second)

9 V battery

p.d. = 9 V = 9 joules transferred
    to each coulomb

power = 2 coulombs per second,
    of 9 joules each
    = 18 joules per second

The potential difference across the battery equals the sum of the potential differences across each lamp.
That is $V = V_1 + V_2$.

Resistance

All components in an electrical circuit have a resistance to current flowing through them. The relationship between voltage, current and resistance in electrical circuits is given by this equation.

$V = I R$

V is the voltage in volts (V)

I is the current in amps (A)

R is the resistance in ohms (Ω)

It is important to be able to rearrange this equation when performing calculations. Use the triangle on the right to help you.

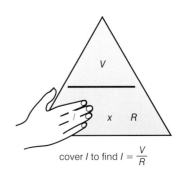

cover I to find $I = \dfrac{V}{R}$

A* EXTRA

- This equation defines the resistance of the component at a certain measured current. For some components, notably for metal wires and resistors, the resistance is fixed and does not change if the current through the device changes. So if the voltage across the device doubles, the current through it doubles.
- If the device has a fixed value of R, then you can calculate V or I if you know the other value. This equation is known as 'Ohm's law', but it's scarcely a law, as it is not true for most materials. It's not even true for metals if their temperature is allowed to change.

In a metal structure metal ions are surrounded by a cloud or 'sea' of electrons.

WORKED EXAMPLES

1   Calculate the resistance of a heater element if the current is 10 A when it is connected to a 230 V supply.

Write down the formula in terms of R: $R = \dfrac{V}{I}$

Substitute the values for V and I: $R = \dfrac{230}{10}$

Work out the answer and write down the unit: $R = 23 \ \Omega$

2   A 6 V supply is applied to 1000 Ω resistor. What current will flow?

Write down the formula in terms of I: $I = \dfrac{V}{R}$

Substitute the values for V and R: $I = \dfrac{6}{1000}$

Work out the answer and write down the unit: $I = 0.006 \ A$

EFFECT OF MATERIAL ON RESISTANCE

Substances that allow an electric current to flow through them are called **conductors**; those which do not are called **insulators**.

Metals are conductors. In a metal structure, the metal atoms exist as positive ions surrounded by an electron cloud. If a potential difference is applied to the metal, the electrons in this cloud are able to move and a current flows.

When the electrons are moving through the metal structure, they bump into the metal ions and this causes **resistance** to the electron flow or current. In different conductors the ease of flow of the electrons is different and so the conductors have different resistances. For instance, copper is a better conductor than iron.

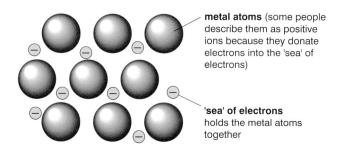

metal atoms (some people describe them as positive ions because they donate electrons into the 'sea' of electrons)

'sea' of electrons holds the metal atoms together

The table below lists materials ranging from the best conductor to the best insulator. The wide range of electrical resistance was one of the great puzzles for Victorian scientists, and it was almost impossible for them to understand how one material can be 1000 000 000 000 000 000 000 000 000 times better at conducting current than another material.

Silver	metal	conductor (best)
Copper	metal	conductor
Aluminium	metal	conductor
Iron	metal	conductor
Graphite		conductor
Soil		
Water		
Silicon		semiconductor
Wood		
Rock		
Most plastics		insulator
Oil		insulator
Glass		insulator
Teflon®		insulator (best)

EFFECTS OF LENGTH AND CROSS-SECTIONAL AREA

For a particular conductor, the resistance is **proportional to length**. The longer the conductor, the further the electrons have to travel, the more likely they are to collide with the metal ions and so the greater the resistance. So a wire that is twice as long will have twice as much resistance.

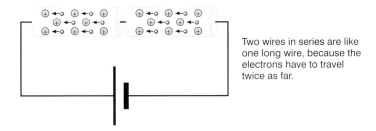

Two wires in series are like one long wire, because the electrons have to travel twice as far.

Resistance is **inversely proportional to cross-sectional area**. The greater the cross-sectional area of the conductor, the more electrons there are available to carry the charge along the conductor's length and so the lower the resistance. So a wire with twice the cross-sectional area, will have half the resistance. (Remember that if the wire is of twice the diameter, then its cross-sectional area will be four times greater, and so the resistance of the wire will be one quarter as much.)

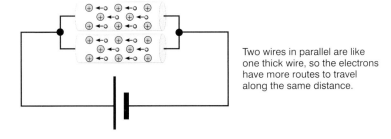

Two wires in parallel are like one thick wire, so the electrons have more routes to travel along the same distance.

You can control the amount of current flowing through a circuit by changing the resistance of the circuit using a **variable resistor** or **rheostat**. Adjustment of the rheostat changes the length of the wire the current is in.

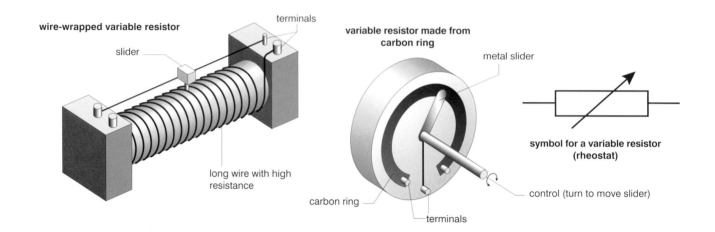

wire-wrapped variable resistor

terminals

slider

long wire with high resistance

variable resistor made from carbon ring

metal slider

carbon ring

terminals

control (turn to move slider)

symbol for a variable resistor (rheostat)

Variable resistors are commonly used in electrical equipment, for example in the speed controls of model racing cars or in volume controls on radios and hi-fi systems.

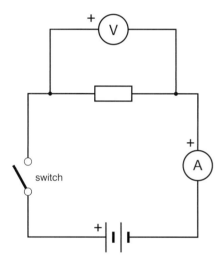

switch

MEASURING RESISTANCE

The resistance of a component can be found using this circuit (left). The component (lamp, resistor or whatever) is placed in a circuit with an ammeter to measure the current through the component, and with a voltmeter to measure the potential difference across it. To power the circuit you could use a battery as shown, or you could use a power supply with a suitable output. To take readings, the circuit is switched on, and readings are made of the p.d. and the current.

The resistance is calculated from the following equation:

$$R = \frac{V}{I}$$

Note that the readings may change a little over the first few seconds. If so, this is probably because the component is heating up and its resistance is changing. If this happens, you would have to decide whether to take the readings before the component has heated up, and so measure the resistance at room temperature, or to wait until the readings have stopped changing. This would give you the 'steady-state' resistance with the component at its usual running temperature.

You may wish to change the e.m.f. of the battery by changing the number of cells (or you may adjust the output of the power supply). If the component is a perfect resistor then you will get the same answer for the resistance; but you will often find that the resistance of the component varies.

EFFECT OF TEMPERATURE ON RESISTANCE

If the resistance of a conductor remains constant, a graph of voltage against current is a **straight line**. The gradient of the line will be the resistance of the conductor.

The resistance of most conductors becomes higher if the temperature of the conductor increases. As the temperature rises, the metal ions vibrate more and provide greater resistance to the flow of the electrons. For example, the resistance of a filament lamp becomes greater as the voltage is increased and the lamp gets hotter.

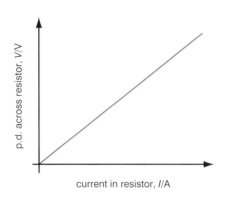

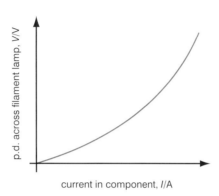

In an 'ohmic' resistor, such as carbon, Ohm's law applies and the voltage is directly proportional to the current – a straight line is obtained. In a filament lamp, Ohm's law is not obeyed because the heating of the lamp changes its resistance.

Electrical energy

All electrical equipment has a **power rating**, which indicates how many joules of energy are supplied each second. The unit of power used is the **watt** (W). Light bulbs often have power ratings of 60 W or 100 W. Electric kettles have ratings of about 2 kilowatts (2 kW = 2000 W). A 2 kW kettle converts 2000 J of energy each second.

The power of a piece of electrical equipment depends on the voltage and the current:

$$P = V I$$

P = power in watts (W)

V = voltage in volts (V)

I = current in amps (A)

cover V to find $V = \dfrac{P}{I}$

WORKED EXAMPLES

1 What is the power of an electric toaster if a current of 7 A is obtained from a 230 V supply?

Write down the formula in terms of P:	$P = V I$
Substitute the values:	$P = 230 \times 7$
Work out the answer and write down the unit:	$P = 1610\,W$

2 An electric oven has a power rating of 2 kW. What current is measured when the oven is used with a 230 V supply?

Write down the formula in terms of I:	$I = \dfrac{P}{V}$
Substitute the values:	$I = \dfrac{2000}{230}$
Work out the answer and write down the unit:	$I = 8.7$ A

CALCULATING THE ENERGY TRANSFERRED

The energy transferred by an appliance depends on the power rating and the time the appliance is running.

$E = Pt$

E = energy in joules (J)
P = power in watts (W)
t = time in seconds (s)

Since power is linked to voltage and current, this equation can also be written as:

$E = VIt$

E = energy in joules (J)
V = voltage in volts (V)
I = current in amps (A)
t = time in seconds (s)

WORKED EXAMPLE

Calculate the energy transferred when a 12 V motor, running at a current of 0.5 A, is left on for 5 minutes.

Write down the formula:	energy = voltage × current × time
Substitute the values: (remember the time *must* be in seconds)	energy = 12 × 0.5 × 300
Work out the answer and write down the unit:	energy = 1800 J

REVIEW QUESTIONS

Q1 **a** A charge of 10 coulombs flows through a motor in 30 seconds. What is the current flowing through the motor?
b A heater uses a current of 10 A. How much charge flows through the lamp in:
i 1 second, ii 1 hour?

Q2 An electric motor drives a water pump that lifts water out of a well that is 10 m deep. It can deliver 360 kg per minute out of the tap at the top.
a How much potential energy is given to the water each second? Hence, what power must be provided by the electric motor? Assume that the motor and pump have 100 per cent efficiency.
b If the motor is designed to run on 12 V, what current will it take out of a 12 V supply when it is working?
c If the motor is designed to work on 220 V, what current will it take out of a 220 V supply when it is working? (NB The 220 V will be an alternating current supply, but this does not affect the calculation.)
d Name one advantage and one disadvantage of the 220 V system over the 12 V system.

Q3 Explain the following observations:
a You rub a plastic pen with a piece of dry kitchen paper, and then put it a few millimetres away from a thin stream of water flowing from a tap. The stream of water is affected by the pen.
b You rub an inflated balloon on a dry piece of cloth and hang it from the ceiling using a piece of sewing-thread. You rub a second balloon on the same piece of cloth, and find that it repels the first balloon.
c The variable resistor used to control a toy racing car stays cold if the car is going at full speed or if the car is stationary. However it gets very hot if the car is going at half speed.

More questions on the CD ROM

Examination questions are on page 156.

3 ELECTRIC CIRCUITS

Circuit diagrams

When people started using electricity, they quickly found that it was not convenient to draw accurate pictures of the circuits that they made. It was much easier to understand how the circuit worked, and to correct faults, if they used standard symbols for the parts. It was also much easier if the wires were drawn in straight lines, rather than trying to copy the exact route taken.

Study the circuits used in this chapter and learn the symbols and what they represent.

THE FLASHLIGHT

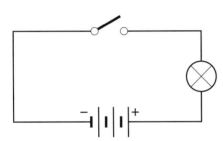

This simple circuit shows how a flashlight is powered by a **battery** consisting of three 1.5 V **cells**, giving a total e.m.f. of 4.5 V. In the case of a flashlight, the cells are put in separately, but in the case of a 9 V battery, for example, the six cells are pre-assembled by the manufacturer. The word 'battery' means an assembly of several cells, but people often use the word to refer to a single cell.

The '+' terminal of the cell is indicated by the long thin line, and the '-' terminal by the short thick line. It may help you to remember this if you imagine yourself cutting the long thin line into two shorter pieces and turning them into a + sign.

The other symbols in the circuit are the **normally open switch**, and the **lamp**.

THE DIODE

It does not matter which way the charge flows through a lamp, but a pocket calculator, say, could be destroyed if the battery is not inserted correctly. One way to prevent this is to add a diode to the circuit.

In this circuit the calculator is represented as a resistor. A calculator is far more complicated than that, but it does behave to the battery *as if* it were a resistor, drawing a small current *I* out of the battery.

As you can see, the arrow on the diode shows the way that conventional current flows. When the battery is inserted the wrong way round, no charge can flow.

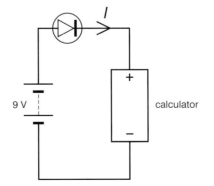

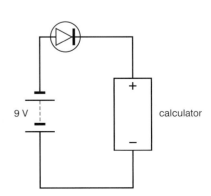

Series and parallel circuits

There are two different ways of connecting two lamps to the same battery. Two very different kinds of circuit can be made. These circuits are called **series** and **parallel** circuits.

	Series	**Parallel**
Circuit diagram		
Appearance of lamps	The brightness or dimness of the lamp depends on the voltage rating and the e.m.f. of the battery. Both lamps appear to have the same brightness, both lamps are dim.	The brightness or dimness of the lamp depends on the voltage rating and the e.m.f. of the battery. Both lamps appear to have the same brightness, both lamps are bright.
Battery	The battery is having a hard time pushing the same charge first through one bulb, then another. This means less charge flows each second, so there is a low current and energy is slowly transferred from the battery.	The battery pushes the charge along two alternative paths. This means more charge can flow around the circuit each second, so energy is quickly transferred from the battery.
Switches	The lamps cannot be switched on and off independently.	The lamps can be switched on and off independently by putting switches in the parallel branches.
Advantages/ disadvantages	A very simple circuit to make. The battery will last longer. If one lamp 'blows' then the circuit is broken so the other one goes out too.	The battery will not last as long. If one lamp 'blows' the other one will keep working.
Examples	Christmas tree lights are often connected in series.	Electric lights in the home are connected in parallel.

COMBINING RESISTORS

Two resistors can be replaced by a single resistor that has the same effect in the circuit. Calculating the value of the resistor needed depends on whether the original resistors are connected in series or parallel.

If the resistors R_1 and R_2 are in ...	Then the combined resistance, R_C, is ...
series	$R_C = R_1 + R_2$
parallel	$R_C = \dfrac{R_1 \times R_2}{R_1 + R_2}$

WORKED EXAMPLES

1 If the two resistors are 1000 ohms and 3000 ohms and they are in series, then

$R_C = (1000 + 3000)$ ohms

   $= 4000$ ohms

$R_1 = 1000\ \Omega$ $R_2 = 3000\ \Omega$ $=$ $R_C = 4000\ \Omega$

Note that the combined resistance is greater than the value of either of the two resistors.

2 If the two resistors are 20 ohms and 30 ohms, and they are in parallel then

$$R_C = \frac{20 \times 30}{(20 + 30)} \text{ ohms}$$

$$= \frac{600}{50} \text{ ohms}$$

$$= 12 \text{ ohms}$$

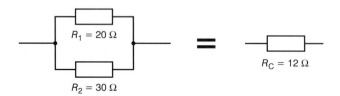

Note that the combined resistance is less than the value of either of the two resistors.

DIRECT AND ALTERNATING CURRENTS

A battery produces a steady current. The electrons are constantly flowing from the negative terminal of the battery round the circuit and back to the positive terminal. This produces a **direct current** (d.c.).

The mains electricity used in the home is quite different. The electrons in the circuit move backwards and forwards. This kind of current is called **alternating current** (a.c.). In some countries mains electricity moves forwards and backwards 50 times each second, that is, with a frequency of 50 hertz (Hz). The frequency chosen varies from country to country.

The advantage of using an a.c. source of electricity rather than a d.c. source is that it can be transmitted from power stations to the home at very high voltages, which reduces the amount of energy that is lost in the overhead cables.

Here is a circuit where the current in and the p.d. across an electric kettle are being monitored.

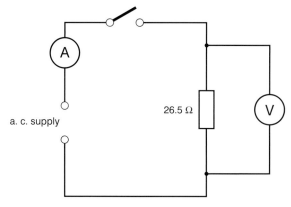

As you can see, we are monitoring the current through the 26.5 ohm heater in the kettle, and the potential difference across it. In this circuit, the p.d. across the heater will have the same value as the e.m.f. of the a.c. supply that it is connected to.

When the normally open switch is closed, the p.d. and the current will have the waveforms shown right. They will both alternate positive and negative at the same time.

To drive the current back and forth, the e.m.f. has to repeatedly change direction. It is almost as if the circuit is being driven by a battery, but the wires to the two terminals on the battery are swapped over a hundred times per second.

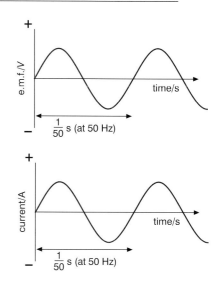

Action and use of circuit components

THE VARIABLE RESISTOR

A variable resistor contains a length of resistance wire and an adjustable sliding contact. One end of the wire and the contact are connected into the circuit. As the contact can be moved from one end of the wire to the other, the resistance of the variable resistor can be set to any value from approximately zero to the whole resistance of the resistance wire inside the device. In the case of the variable resistor here, the value can be set to any value between 0 ohms and 500 ohms.

In this circuit, we are using a variable resistor to control the speed of an electric motor. If it is a 24 V electric motor, and a battery with e.m.f. of 24 V, then with the variable resistor set to 0 ohms, the p.d. across the motor will be 24 V, and the motor will run at full speed. The p.d. across the variable resistor will be 0 V. The resistance of the whole circuit is 12 ohms and so the current through the motor will be 2.0 A.

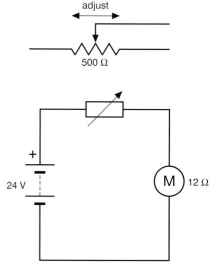

If the variable resistor is set to 12 ohms, then the total resistance in the circuit is now 24 ohms, and the total current can be calculated:

$$I = \frac{V}{R}$$
$$= \frac{24}{24}$$
$$= 1.0 \text{ A}$$

The current through the motor will have halved, and the motor will run slower. Note that we can now work out the p.d. across the motor.

current through the motor = 1.0 A

its resistance = 12 ohms

$$\text{p.d.} = I \times R$$
$$= 1.0 \times 12$$
$$= 12 \text{ V}$$

Likewise, the p.d. across the resistor is 12 V, and the p.d. values around the circuit add up to 24 V, which is the same as the e.m.f. of the battery, as always.

THE POTENTIOMETER AND THE POTENTIAL DIVIDER

The potentiometer is very similar in its design to a variable resistor, and in fact the same component can usually be used as either device. In the potentiometer, both ends of the resistance wire and the adjustable contact, all three points, are connected into the circuit. The two ends of the resistance wire are connected to both ends of the battery or power supply. So if the battery is 5 V, then the p.d. across the potentiometer is 5 V. If the slider is set to the top, then the p.d. V_{out} across the two output wires will again be 5 V. But if the slider is set to minimum, the two output wires are connected to the same point and the p.d. will be 0 V. This is a major difference between a potentiometer and a variable resistor: the output of the potentiometer can be set to zero. This is one reason why the volume control on most audio equipment is a potentiometer, as it gives full control over the output volume.

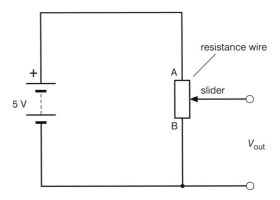

The potentiometer acts as a potential divider: it divides up the p.d. supplied by the power source.

It is also possible to make a potential divider using two resistors as shown in this circuit.

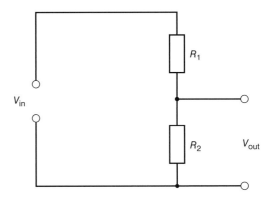

If $R_1 = R_2$, then the output p.d. V_{out} will be just half the input p.d. V_{in}. Note that it does not really matter what value R_1 and R_2 have; it is more important to note their *relative* values. So if R_1 is very small compared with R_2, then the output will be high, it will be approaching the value of V_{in}. If R_2 is very small compared with R_1, then the output will be close to zero, because the two output wires will almost be joined together.

THE THERMISTOR AND THE LDR

In some substances, increasing the temperature actually **lowers** the resistance. This is the case with **semiconductors** such as silicon. Silicon has few free electrons and so behaves more like an insulator than a conductor. But if silicon is heated, more electrons are removed from the outer electron shells of the atoms producing an increased electron cloud. The released electrons can move throughout the structure, creating an electric current. This effect is large enough to outweigh the increase in resistance that might be expected from the increased movement of the silicon ions in the structure as the temperature increases.

Semiconductors are used to make **thermistors**, which are used as temperature sensors, and **light–dependent resistors** (LDRs), which are used as light sensors.

In LDRs it is light energy that removes electrons from the semiconductor atoms, increasing the electron cloud.

INPUT TRANSDUCERS

A transducer is a device that transfers energy from one form to another. An input transducer may be needed to transfer, for example, heat or light energy to electrical energy. Many machines have to take action if exposed to light or heat or some other input. The controls of a boiler will light the flame if the water is too cold; the computer controlling the house will close the curtains when it gets dark. The LDR and the thermistor are suitable electrical transducers for giving an electrical signal for this purpose. They are best used as part of a potentiometer.

In this circuit (right), the LDR has a resistance of about 200 kΩ in the dark. This means that R_2 is much smaller than R_1, and the output will be small, (about 0.2 V). When the LDR is in the light, its resistance drops to about 3 kΩ, and so the output voltage rises to something near to the input voltage (about 3.8 V).

So with this circuit, light is detected by the output going from **low** to **high**. If the LDR had been put in the lower position, with a resistor above, then light would be detected by the output going from high to low.

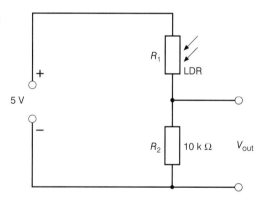

THE RELAY

The relay is an electromagnet that can operate one or more switch contacts. For example, the contacts in this relay (right) join points A and B when the switch is open. When the electromagnet is energised it attracts a piece of soft iron and joins points A and C. Points B and C are never joined. A relay like this can be used as an output transducer, as you can choose a relay that operates at a particular voltage. The relay transfers electrical energy to mechanical energy.

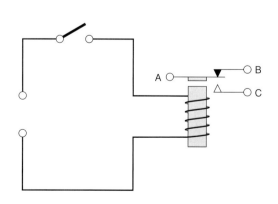

Another application for the relay is to allow you to use a

small current in the control circuit to close the contacts for a very large current in a second circuit. For example the starter motor in a car is connected to the battery by wires that are about 10 mm in diameter, so high is the current that flows to the starter motor. It would be completely impractical to have wires of this size to the switch operated by the car's key, so the key controls only the current to the coil, and the relay contacts control the current to the motor.

A relay circuit used to switch on a starter motor.

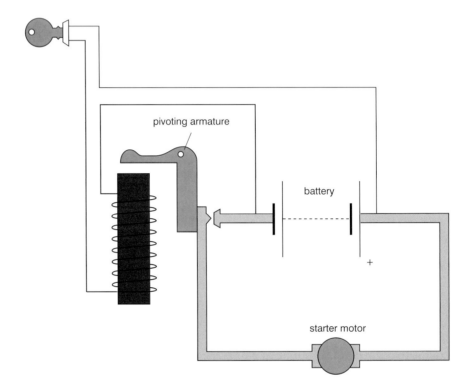

THE CAPACITOR

The capacitor is designed to store electric charge (and hence electrical energy) temporarily. The energy is stored as an electric field between two plates. The space between the plates is filled with an insulator, so current cannot flow through a capacitor. In some applications it is only stored for a fraction of a second, but be warned that large capacitors can hold on for many days to quantities of electricity that can kill. The following experiment must not be done with more than 12 V.

The following experiment shows how the capacitor can store energy.

In the first step (below) connect a 12 V d.c. power supply to the capacitor.

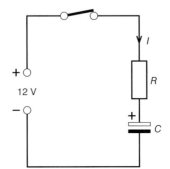

Note that many capacitors are marked with + and – terminals. The symbol for a capacitor of this type has a rectangular box shape for the + terminal. If the terminals are connected the wrong way round the capacitor will be damaged or destroyed. Capacitors also have a voltage rating. This one must be rated for 12 V or higher.

When, in the first step, the switch is closed, current flows *into* the capacitor as shown. A resistor R should be fitted to prevent too strong a surge of current. Initially, the p.d. across the capacitor is 0 V, and the full 12 V is across the resistor. So if the resistor is 12 ohms, for example, the initial current will be 1 A. As the capacitor stores charge, the p.d. across it increases to 12 V. If the capacitor has a large value, then the current will flow for longer and more energy will be stored. Remember that a current of 1 amp for 1 second into the capacitor will store a charge of 1 coulomb. When it is fully charged, the p.d. across the capacitor will be 12 V, the p.d. across the resistor will be 0 V, and the current will have died to zero.

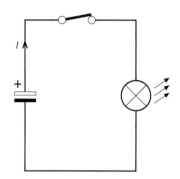

In the second step (right) the charged capacitor is connected to a 12 V lamp. Then when the switch is closed, the lamp will have 12 V of p.d. across it, and it will be lit with maximum brightness as the current flows out of the capacitor. But the voltage of the capacitor will steadily drop and the lamp will get dimmer and dimmer until the current stops completely.

The capacitor can be used in a time-delay circuit.

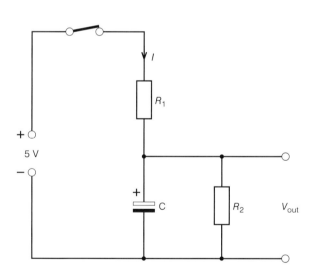

In this circuit, when the switch is closed, the p.d. across the capacitor C will increase from zero. If the output is connected to an output transducer, such as a relay, that operates when the output voltage reaches, say, 3 V, then there will be a time delay between the switch closing and the relay operating. This delay could be a fraction of a second or many minutes, depending on the resistance of R_1 and the capacitance of the capacitor C. For a long delay you want a high value of R_1 (to give a low current I) and a large capacitor C (to make it charge up more slowly).

You may need to add a resistor R_2 so that the p.d. across the output is returned to zero when the switch is opened. R_2 must have a high resistance, or the capacitor will never be able to charge up (because the current would just flow through R_2 instead).

THE DIODE AS RECTIFIER

If you change the power source in the circuit on page 131 to a.c., the current will keep on changing direction. If a diode is added to the a.c circuit then, although the e.m.f. will stay as before, current will only flow in one direction. Conventional current can flow in the direction of the arrow in the diode symbol.

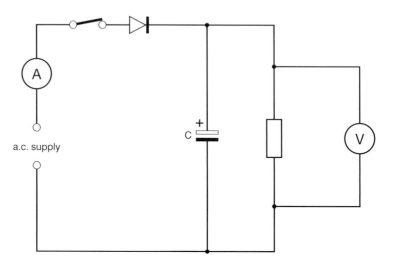

a.c. supply

The graphs below show what you will measure across the resistor if the capacitor is *not* fitted. Only the positive p.d. will appear across the resistor. We say that the diode has rectified the alternating current. However, the p.d. and current fluctuate – they do not have a steady value.

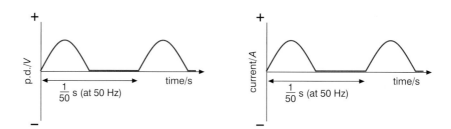

If you fit a capacitor C as well, then you have in effect a d.c. power supply. Note that the capacitor must be fitted this way round. The capacitor will be charged up by the pulses of electricity, and will be able to deliver a steady forward current through the resistor even when the e.m.f. is backwards.

THE TRANSISTOR AS SWITCH

The transistor is another output transducer. It is a semiconductor device made of silicon. It is the building block of electronics, and the processor of a computer contains millions of them. Most transistors are small and run at a few volts, but modern electric trains are started and stopped by semiconductor devices that control thousands of volts and thousands of amps.

The transistor has three wires coming out of it. These are labelled:
b = base
c = collector
e = emitter.

This type of transistor is known as a NPN transistor, with the arrow pointing out of the emitter.

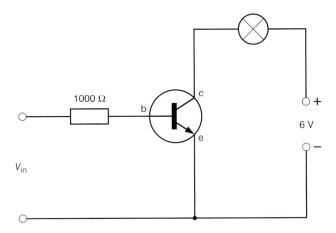

There are really two circuits here, one with the base and the emitter, which are in the input circuit, and one with the collector and the emitter, which are in the output circuit. (This overall circuit is known as a common-emitter circuit.)

In this circuit, if no voltage is applied to the base, or if the two input connections are joined together, then almost no current will flow through the output circuit, and the lamp will not light. If the input to the base is raised above 0.6 V, then the lamp will switch on. The 1000 ohm resistor is there to protect the input of the transistor, and allows the input to be set higher than 0.6 V, to 5 V or more without harming the transistor. Very little current is needed in the input circuit (and some transistors require no current at all). The output circuit can handle much higher currents, the exact current depending on the transistor chosen.

LIGHT-SENSITIVE SWITCH

Here is the circuit of the light-sensitive switch, using a relay output.

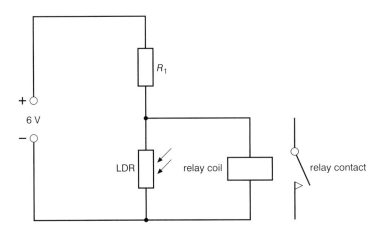

The relay is not energised so long as the resistance of the LDR stays low. If the illumination reduces, the resistance of the LDR increases and the p.d. across the coil goes up. When the illumination is sufficiently low, the relay will close the contact. Note that the circuit controlled by the relay does not need to have any connection whatsoever to the relay coil circuit. This can be an important safety feature of the circuit.

TEMPERATURE-OPERATED ALARM

In this circuit, the transistor will start to conduct electricity, and the bell will sound, if the p.d. across resistor R_2 goes to more than 0.6 V. The value of R_2 must be set to make this occur at the correct temperature. If R_2 was made a variable resistor, then the temperature of the alarm could be adjusted.

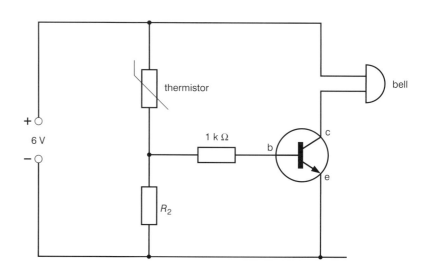

Some DJs still prefer vinyl records. If you look at the record you can see the analogue signal. The loudspeaker cone accurately follows these sideways movements of the groove.

Each DVD contains about 4.7 gigabytes of data in the form of the numbers 1 and 0. If the numbers out of one DVD were printed in paper books, you would need over 3000 books, each of 1000 pages.

Digital electronics

In digital electronics, the circuit is only allowed to be in one of two states: **on** or **off**. These two states represent the numbers 1 and 0. In practice, to make the circuit reliable, any low voltage is taken to indicate zero, and any high voltage is taken to represent 1. Information is sent from one place to another as a long stream of 1 and 0 numbers. Each number is known as a bit. Computers and DVD players use digital electronics. A DVD player has to read about 5 million bits per second to display the movie picture on the screen.

In analogue electronics, as is used by an amplifier driving a loudspeaker, the current through the loudspeaker varies in a complex way that accurately describes the way that the loudspeaker should move. All of the human senses are analogue, and so there must always be analogue-to-digital converters between man and machine. Until the 1980s, equipment used to be entirely analogue, and some enthusiasts claim, with some evidence, that vinyl records, recorded and played back using analogue equipment, sound best. However, most equipment is now digital, as it gives good quality and the ability to add many extra features at low cost.

WHAT ARE LOGIC GATES?

A logic gate is an electronic circuit that has one or more **input** signals and one **output** signal. These signals are voltages that can be HIGH (about 5 V) or LOW (about 0 V). Logic gates are **digital** circuits as they can only have certain values of input and output – high or low. The output signal depends on the combination of signals at the inputs.

WHAT ARE TRUTH TABLES?

Truth tables summarise the way in which a logic gate operates. Truth tables usually use 1 for a HIGH signal and 0 for a LOW signal.

AND gate
The output is high only when input A **AND** input B are high.

Inputs A	B	Output
0	0	0
0	1	0
1	0	0
1	1	1

OR gate
The output is high when input A **OR** input B is high, OR both.

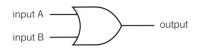

Inputs A	B	Output
0	0	0
0	1	1
1	0	1
1	1	1

NOT gate
This gate is also called an **inverter**. It has only one input.
The output is high when the input is **NOT** high.

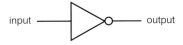

Input	Output
0	1
1	0

WORKED EXAMPLES

1 For safety, a car engine will not start unless the door is closed and the seat belt is fastened. Which type of logic gate is needed?

For the engine to start, both input conditions must be met. This circuit will need an AND gate.

2 A doorbell has switches at the front door and the back door of a house. Which type of logic gate is needed?

The bell needs to ring if either switch is pressed (or both). This circuit will need an OR gate.

MORE LOGIC GATES

A **NOR** gate combines an OR gate and a NOT gate.
The output is high when *neither* A NOR B are high.

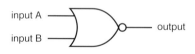

Inputs A	B	Output
0	0	1
0	1	0
1	0	0
1	1	0

A **NAND** gate combines an AND gate and a NOT gate.
The output is high when *both* inputs are *not* high.

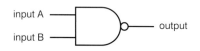

Inputs A	B	Output
0	0	1
0	1	1
1	0	1
1	1	0

LOGIC GATE CIRCUITS

Logic gates can be combined – the output signal from one gate can be used
as the input signal to another.

WORKED EXAMPLE

Draw the truth table for this combination.

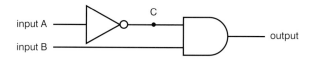

The key is to mark C on the diagram and work out this value first.

The value at C is NOT input A.

Inputs A	B	C	Output
0	0	1	
0	1	1	
1	0	0	
1	1	0	

The final output is input B AND
the value at C.

Inputs A	B	C	Output
0	0	1	0
0	1	1	1
1	0	0	0
1	1	0	0

BISTABLE CIRCUITS

These circuits are made by cross-linking NOR gates or NAND gates.

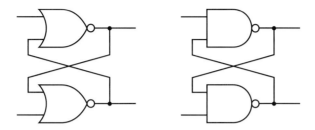

The output depends on the **sequence** of changes at the inputs – the circuits act as a simple 'memory'. If only one output is used, the circuit is called a **latch**.

In a NOR gate latch:

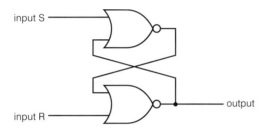

- Initially, both inputs are low and the output is low.
- Input S (the **set** input) goes high, so the output goes high.
- Input S returns to low, but the output stays high – it is **latched on**.
- Input R (the **reset** input) goes high, so the output returns to low.
- Input R returns to low and the output stays low – it is **latched off**.

This is useful in circuits such as burglar alarms where the input sensor may only send a signal for a short time. The latch circuit keeps the alarm on until the reset is used.

In a NAND gate latch, the sequence is the same except that:
- Both inputs and the output are *high* at the start.
- Moving the set input briefly to *low* changes the output to low.
- Moving the reset input briefly to *low* returns the output to high.

HOW DO WE PROVIDE SIGNALS FOR LOGIC GATES?

The simplest way to provide an input signal for a logic gate is to use a switch and a resistor.

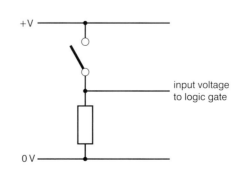

Switch	Input voltage to logic gate
open	LOW
closed	HIGH

SENSORS USING POTENTIAL DIVIDERS

A sensor is a circuit in which a voltage changes as the environment changes. Most sensor circuits use a **potential divider**.

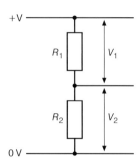

CALCULATING VOLTAGES IN A POTENTIAL DIVIDER

The current is the same in both resistors, so:

$$\frac{V_1}{V_2} = \frac{R_1}{R_2}$$

Often, we only need to work out the input voltage to the logic gate.

In this case, we can use the equation in this form:

$$V_{out} = \left(\frac{R_2}{R_1 + R_2} \right) V$$

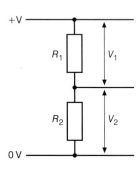

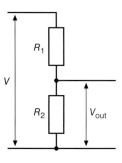

USING A LIGHT-DEPENDENT RESISTOR (LDR)

The resistance of an **LDR** is high in the dark and low in the light.

When it is *light*, the LDR has a *low* resistance, so it has a small share of the voltage. The input voltage to the logic gate is high.

When it is *dark*, the LDR has a *high* resistance, so it has a large share of the voltage. The input voltage to the logic gate is low.

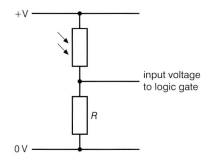

This time the resistors have been switched round. The input voltage to the logic gate will be high when it is dark and low when it is light.

If the resistor R is changed to a variable resistor, then we can vary the light level at which the input voltage to the logic gate goes from low to high.

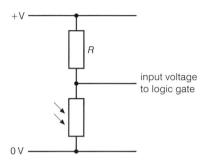

USING A THERMISTOR

The resistance of a **thermistor** is high at low temperatures and low at high temperatures.

When it is *hot*, the thermistor has a *low* resistance, so it has a small share of the voltage. The input voltage to the logic gate is high.

When it is *cold*, the thermistor has a *high* resistance, so it has a large share of the voltage. The input voltage to the logic gate is low.

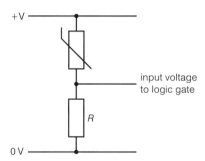

This time the resistors have been switched round. The input voltage to the logic gate will be high when it is cold and low when it is hot.

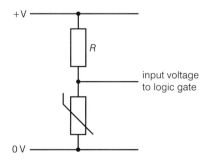

OTHER SENSORS

A **moisture sensor** has two wire probes separated by a small gap.

Moisture (water) conducts electricity better than the air, so the resistance of the gap changes when it is wet. This changes the input voltage to the logic gate.

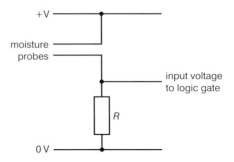

In a **tilt switch**, two contacts are only connected when the switch is turned in a particular direction. This changes the resistance, varying the input voltage to the logic gate.

Other sensors include:
- **magnetic sensors** – a change in magnetic field causes a change in the input voltage to the logic gate
- **pressure sensors** – the resistance of the sensor changes when the sensor is pressed in some way.

OUTPUT DEVICES

The output signal from a logic gate is large enough to drive devices such as **buzzers** and **LEDs (light–emitting diodes)**. When an LED is used, a resistor is used in series with it. This **protects** the LED against the current being too high.

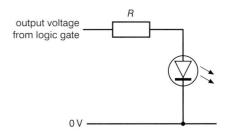

The LED will light when the output from the logic gate is high (5 V).

WORKED EXAMPLE

When lit, an LED has a voltage of 1.8 V across it and a current of 10 mA. Work out a suitable value for the protective resistor, *R*.

voltage across resistor = output voltage from logic gate – voltage across LED

$$= 5\,V - 1.8\,V$$

$$= 3.2\,V$$

Current through resistor = 10 mA = 0.01 A (because resistor and LED are in series)

Write Ohm's law in terms of *R*: $R = \dfrac{V}{I}$

Substitute the values for *V* and *I*: $R = \dfrac{3.2}{0.01}$

Work out the answer and write down the unit: $R = 320\ \Omega$

REVIEW QUESTIONS

Q1 Look at the circuit diagrams on the right. They show a number of ammeters and in some cases the readings on these ammeters. All the lamps are identical.

  a For circuit X, what readings would you expect on ammeters A_1 and A_2?

  b For circuit Y, what readings would you expect on ammeters A_4 and A_5?

  c Look at the last circuit diagram. It shows how three voltmeters have been added to the circuit. What reading would you expect on V_1?

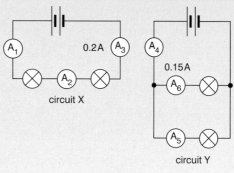

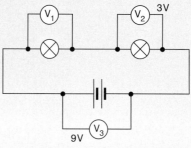

Q2 Complete the truth tables to show that:

  a a NOR gate is the same as an OR gate followed by a NOT gate.

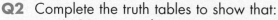

Inputs			Output
A	**B**	**C**	
0	0		
0	1		
1	0		
1	1		

  b a NAND gate is the same as an AND gate followed by a NOT gate.

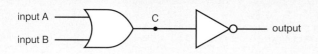

Inputs			Output
A	**B**	**C**	
0	0		
0	1		
1	0		
1	1		

Q3 Complete the truth table for this circuit.

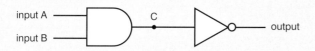

Inputs			Output
A	**B**	**C**	
0	0	0	
0	0	1	
0	1	0	
0	1	1	
1	0	0	
1	0	1	
1	1	0	
1	1	1	

Q4 Here is a light sensor circuit. It has to detect light and dark.

  a At a particular light level, the resistance of the LDR is 500 Ω. Calculate the input voltage to the logic gate.

  b In the dark the resistance of the LDR is 10 KΩ. What is the input voltage now?

  c State two reasons why a relay must be used when the output from the logic gate needs to control a room heater.

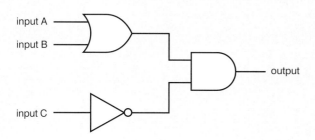

Examination questions are on page 156.

More questions on the CD ROM

4 Dangers of electricity

Electricity is an effective method for transferring energy in many domestic situations. Often, electric circuits are used to transfer energy to movement (using an electric motor) or to heating. Electrical heating can be used in cooking and in heating a building (for example, through fan heaters or radiators).

Electricity can cause hazards in domestic situations. For example:

Hazard	Danger
Frayed cables	wiring can become exposed
Long cables	cause a trip or a fall
Damaged plugs	wiring can become exposed
Water around sockets	water conducts, so can connect a person into the mains supply
Pushing metal objects into sockets	connects the holder to the mains supply

If there is a fault in an electrical appliance, it could take too much electrical current. This might make the appliance itself dangerous, or it could cause the flex between the appliance and the wall to become too hot and start a fire.

Electrical appliances can be damaged if the current flowing through them is too high. The electric current usually has to pass through a **fuse** or circuit breaker before it reaches the appliance. If there is a sudden surge in the current, the wire in the fuse will heat up and melt – it 'blows'. This breaks the circuit and stops any further current flowing. If a circuit breaker is used, then the circuit breaker springs open (trips) a switch if there is an increase in current in the circuit. This can be reset easily after the fault in the circuit has been corrected.

In many houses there will be a distribution box that takes all of the electricity for the house and sends it to the different rooms. In old houses this box may still use fuses, but in modern installations, the box will use miniature circuit breakers, often known as MCBs.

Where a fuse is fitted to the plug, it must have a value above the normal current that the appliance needs but should be as small as possible. The most common fuses for plugs are rated at 3 A, 5 A and 13 A.

WORKED EXAMPLES

1 What fuse should be fitted in the plug of a 2.2 kW electric kettle used with a supply voltage of 230 V?

Calculate the normal current: $I = \dfrac{P}{V}$

$$= \frac{2200\,\text{W}}{230\,\text{V}}$$

$$= 9.6\,\text{A}$$

Choose the fuse with the smallest rating bigger than the normal current: the fuse must be 13 A.

2 What fuse should be fitted to the plug of a reading lamp which has a 60 W lamp and a supply of 230 V?

Calculate the normal current: $I = \dfrac{P}{V}$

$$= \dfrac{60\,\text{W}}{230\,\text{V}}$$

$$= 0.26\,\text{A}$$

Choose the fuse with the smallest rating bigger than the normal current: the fuse must be 3 A.

OTHER SAFETY MEASURES

Metal-cased appliances must have an **earth wire** as well as a fuse. If the live wire worked loose and came into contact with the metal casing, the casing would become live and the user could be electrocuted. The earth wire provides a very low resistance route to the 0 V earth – usually water pipes buried deep underground. This low resistance means that a large current passes from the live wire to earth, causing the fuse to melt and break the circuit.

Appliances that are made with plastic casing do not need an earth wire. The plastic is an insulator and so can never become live. Appliances like this are said to be **double insulated**.

In situations that may expose people to electricity unexpectedly, for example using an electric drill, especially drilling into a wall with hidden power cables, or using power tools out of doors, perhaps in wet conditions, a residual current circuit breaker (RCCB) must be used in the power socket on the wall. If any of the charges starts to leak out, the RCCB will turn off the power in 30 ms or less. This may or may not be quick enough to save the user's life.

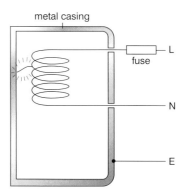

The earth wire and fuse work together to make sure that the metal outer casing of this appliance can never become live and electrocute someone.

REVIEW QUESTIONS

Q1 **a** A hairdryer works on mains electricity of 230 V and takes a current of 4 A. Calculate the power of the hairdryer.
  b In some countries it is illegal to have power sockets in a bathroom, to stop you using hairdryers. Why would it be foolish to use a hairdryer near to a washbasin?

Q2 In her living-room, Felicity has the following items:
  • three 100 W lamps
  • a TV that takes 2 A
  • a hi-fi audio system that takes 1 A
  • a 2 kW electric heater
  • a 3 kW air conditioning unit.
  The whole room is supplied from a 220 V a.c. power supply through one miniature circuit breaker (MCB). What rating of MCB should you fit, if values of 10 A, 20 A, 30 A, 40 A, 50 A and 60 A are available? How would your answer change if the supply were 120 V a.c.?

More questions on the CD ROM

Examination questions are on page 156.

5 ELECTROMAGNETIC EFFECTS

Videos & questions on the CD ROM

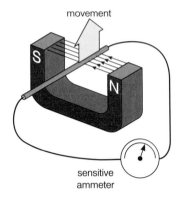
movement

S

N

sensitive ammeter

Electromagnetic induction

Michael Faraday was the first person to generate electricity from a magnetic field using **electromagnetic induction**. The large generators in power stations generate the electricity we need using this process.

Current is created in a wire when:
* the wire is moved through a magnetic field ('cutting' the field lines)
* the magnetic field is moved past the wire (again 'cutting' the field lines)
* the magnetic field around the wire changes strength.

Current created in this way is said to be **induced**.

The faster these changes, the larger the current.

In practice, the changes are induced in a coil of wire, because the current created is increased by the number of turns of wire in the coil. The wire used must not be too thin as carrying the current will then cause it to overheat.

Note that the induced current will flow in a direction that opposes the movement of the wire. See the section on the force on a current-carrying conductor in a magnetic field (page 149).

DYNAMOS AND GENERATORS

A **dynamo** is a simple current generator. It looks very much like an electric motor. Turning the permanent magnet reverses the magnetism through the coil every time the magnet is rotated by 180 degrees. The changes in the magnetic flux through the coil induce an alternating current in the wires. The frequency of the electricity depends on the speed of the bicycle.

In a bicycle dynamo, the magnet rotates and the coil is fixed.

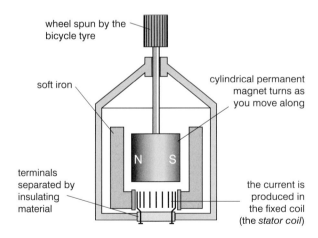

wheel spun by the bicycle tyre

soft iron

cylindrical permanent magnet turns as you move along

N   S

terminals separated by insulating material

the current is produced in the fixed coil (the *stator coil*)

A.C. generator

Power station generators produce **alternating current**. Power stations use electromagnets rather than permanent magnets to create the magnetic field, and then pass the magnetic field through the rotating coils. The generator rotates at a fixed rate, producing a.c. at 50 hertz or 60 hertz, depending on the country.

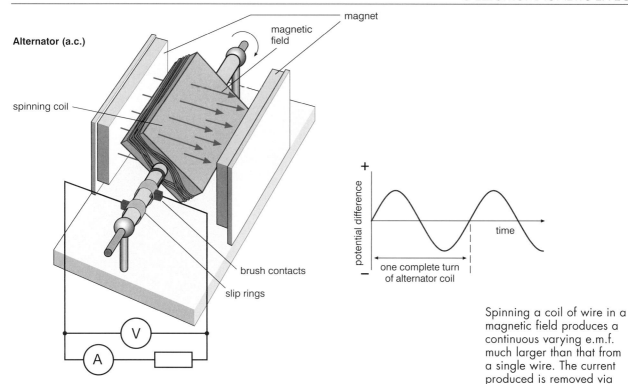

Alternator (a.c.)

magnet

magnetic field

spinning coil

brush contacts

slip rings

one complete turn of alternator coil

potential difference

time

Spinning a coil of wire in a magnetic field produces a continuous varying e.m.f. much larger than that from a single wire. The current produced is removed via slip rings. The output is an alternating current.

Transformer

A **transformer** consists of two coils of insulated wire wound on a piece of iron. If an alternating voltage is applied to the first (primary) coil, the alternating current produces a changing magnetic field in the core. This changing magnetic field induces an alternating current in the second (the secondary) coil.

If there are more turns on the secondary coil than on the primary coil, then the voltage in the secondary coil will be greater than the voltage in the primary coil. The exact relationship between turns and voltage is:

$$\frac{\text{primary coil voltage } (V_\text{p})}{\text{secondary coil voltage } (V_\text{s})} = \frac{\text{number of primary turns } (N_\text{p})}{\text{number of secondary turns } (N_\text{s})}$$

When the secondary coil has more turns than the primary coil, the voltage increases in the same proportion. This is a **step-up transformer**.

A transformer with fewer turns on the secondary coil than on the primary coil is a **step-down transformer**, which produces a smaller voltage in the secondary coil.

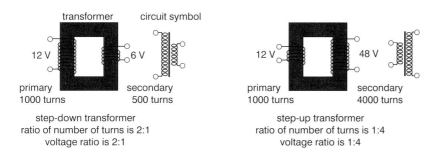

transformer   circuit symbol

12 V   6 V

primary
1000 turns

secondary
500 turns

step-down transformer
ratio of number of turns is 2:1
voltage ratio is 2:1

12 V   48 V

primary
1000 turns

secondary
4000 turns

step-up transformer
ratio of number of turns is 1:4
voltage ratio is 1:4

Transformers are widely used to change voltages. They are frequently used in the home to step down the mains voltage of 230 V to 6 V or 12 V.

WORKED EXAMPLE

Calculate the output voltage from a transformer when the input voltage is 230 V and the number of turns on the primary coil is 2000 and the number of turns on the secondary coil is 100.

Write down the formula:	$\dfrac{V_p}{V_s} = \dfrac{N_p}{N_s}$
Substitute the values known:	$\dfrac{230}{V_s} = \dfrac{2000}{100} = 20$
Rewrite this so that V_s is the subject:	$V_s = \dfrac{230}{20}$
Work out the answer and write down the unit:	$V_s = 11.5\,\text{V}$

The current used by the transformer must change as well. No transformer is 100 per cent efficient, because all transformers produce some heat when they are working. But if it were 100 per cent efficient, then the electrical power going in would equal the electrical power going out. That is to say:

primary coil voltage (V_p) × primary coil current (I_p) = secondary coil voltage (V_s) × secondary coil current (I_s)

For example, if the output is 12 V, 10 A, that is 120 watts of power going out of the transformer. If you know that the input e.m.f. is 240, then the input current will be 0.5 A.

TRANSMITTING ELECTRICITY

Most power stations **burn fuel** to heat water into high-pressure steam, which is then used to drive a **turbine**. The turbine turns an a.c. generator, which produces the electricity.

The most common fuels used in power stations are still coal, oil and gas.

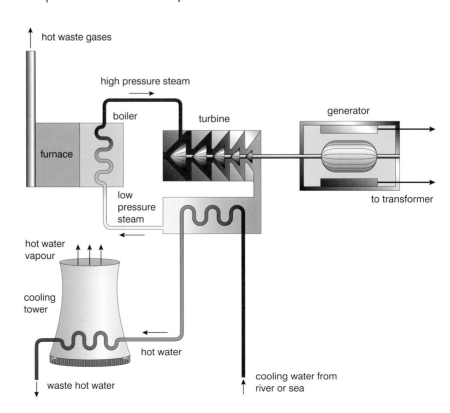

To minimise the power loss in transmitting electricity, the current has to be kept as low as possible. The higher the current, the more the transmission wires will be heated by the current and the more energy is wasted as heat.

This is where transformers are useful. This is also the reason that mains electricity is generated as alternating current. When a transformer steps up a voltage, it also steps down the current and vice versa. Power stations generate electricity with a voltage of 25 000 V. Before this is transmitted, it is converted by a step-up transformer to 400 000 V. This is then reduced by a series of step-down transformers to 230 V before it is supplied to homes.

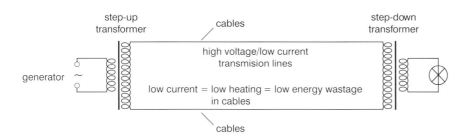

Mains electricity is a.c. so that it can be easily stepped up and down. High-voltage/low-current transmission lines waste less energy than low-voltage/high-current lines.

ENERGY LOSSES IN CABLES

With the exception of some lengths of superconducting cable (which has zero resistance but needs to be kept at a temperature below −200 °C) the distribution cables used by the electricity companies do not have zero resistance. A typical cable with a length of 100 km may have a resistance of 4 ohms. Now consider the problem facing the company when it wants to send 4 MW of power to a town 100 km away. It must send either 10 A at 400 000 V, or 160 A at 25 000 V, or 17 400 A at 230 V.

Well, the 230 V solution is completely hopeless. To send 17 400 A through a resistor of 4 ohms requires a p.d. across the wire of 68 000 V. So almost all of the power from the power station would be used in the cables.

At 25 000 V, the p.d. across the cable would be:
$V = I \times R$
$\quad = 160 \times 4$
$\quad = 640$ V

The power lost in the cables would be:
$P = V \times I$
$\quad = 640 \times 160$
$\quad = 102\ 400$ W

Of the 4 000 000 W being sent, this is 2.6 per cent. This is not too bad, as electricity supply companies expect to lose a total of 5–10 per cent of the power that they generate between the power station and the customer.

At 400 000 V, 10 A, the power lost in the cables is just 400 W, which is 0.01 per cent of the power being sent. These cables will cost more, and so the electricity company will have to work out which high voltage solution is best.

This 400 000 V distribution power line absorbs very little of the power that it carries, but it cost perhaps US$ 500 000 per km to build.

The magnetic effect of a current

If a wire is carrying electric current it generates a magnetic field around itself. The higher the current, the stronger the field. Some people believe that this field is a health hazard, particularly around high voltage distribution lines, but research into the topic has been unable to demonstrate any risk so far.

If the current is travelling along a long straight wire the field looks like this (below).

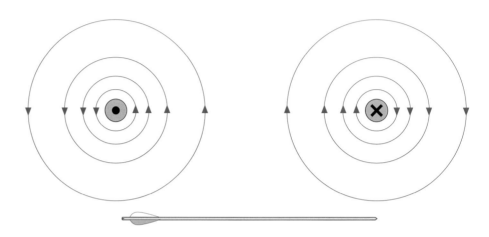

The dot in the centre of the wire indicates that the current is travelling directly towards you; the 'x' indicates that the current is travelling directly away. To remember this, think of an arrow. The dot is the tip of the arrow coming towards you, the 'x' is the flights on the tail of the arrow.

The field lines form continuous rings around the wire all along its length. The lines are shown closest together near to the wire, because the field is strongest there, and quickly gets smaller further away from the wire.

If the current is travelling towards you, the magnetic field lines are going in an anticlockwise direction, and if away from you they are going clockwise. To remember this, think of a woodscrew or a corkscrew. In both cases, if the screw is travelling away from you it is going clockwise.

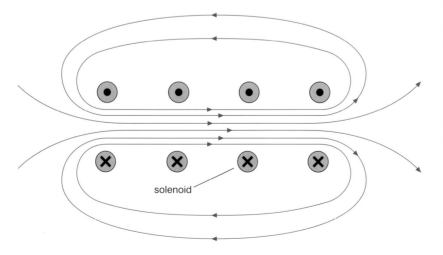

solenoid

We are using the **conventional current**, so remember that the electrons are *actually* going in the opposite direction.

As you saw on page 110, an electric current flowing through a coil of wire (a **solenoid**) creates a magnetic field that looks very similar to the field from a bar magnet. If we were to cut a cross-section through a solenoid, it would look like the diagram (left). The current through the wires is marked as above.

With the current flowing as shown, the magnetic field lines are coming out of the right-hand end of the solenoid, and this is the north pole of the solenoid. To identify which end is the north pole and which end is the south pole, look directly at the end of the magnet and see which way the conventional current is circulating. There is an easy way to remember which is which: the direction arrows of the current can be incorporated into an 'N' or an 'S' (see below).

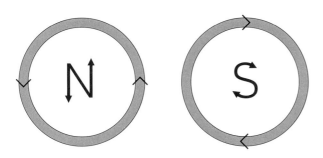

To make the field stronger, you need more turns in the solenoid, and more current through the turns. (And adding a soft iron core makes a big difference as well.) If you reverse the current, the N and S poles will change ends.

Force on a current-carrying conductor

If a wire carrying an electric current passes through a magnetic field, with the field at right-angles to the wire, then the wire will experience a sideways force at right-angles both to the wire and to the magnetic field.

Convince yourself that Fleming's left-hand rule gives you the correct answer for this diagram. Remember that the current is, as usual, the conventional current, and that the electrons are travelling the other way.

The size of the force depends on the magnitude of the current and the strength of the magnetic field. If you experiment with Fleming's left-hand rule you should be able to confirm that if you reverse either the magnetic field or the current then the force will be applied in the opposite direction, but that if you reverse *both* the field *and* the current then the force stays unchanged.

It is useful to look at the magnetic field lines for this set-up (below).

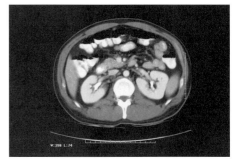

The magnetic field inside the solenoid is remarkably uniform, and this is used in the MRI scanner to allow doctors to produce images of the inside of the body. To do this the whole body has to be placed inside the solenoid.

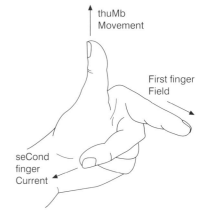

Fleming's left-hand rule predicts the direction of the force on a current-carrying wire.

The field lines from the magnet are dragged to one side by the direction of the field lines that are around the wire. If you imagine that the lines are made of stretched elastic, then it is clear why the wire feels a sideways force.

THE DIRECTION OF AN INDUCED E.M.F.

We may now return to the point made at the beginning of this chapter (page 144). Take the wire and the magnet described in the previous paragraph, with no current flowing. If the wire is moved upwards in the magnetic field, then a current will flow in the wire (if the circuit is complete). The current in the wire will flow towards you, which will give you a resultant force downwards. So as the wire is moved upwards, it will resist you by generating a force downwards. Even if a current does not flow, an e.m.f. will be induced in the wire that will try to generate this current.

More generally 'the direction of an induced e.m.f. opposes the changes causing it'.

D.C. motor

An **electric motor** transfers electrical energy to kinetic energy. It is made from a coil of wire positioned between the poles of two permanent magnets. When a current flows through the coil of wire, it creates a magnetic field, which interacts with the magnetic field produced by the two permanent magnets. The two fields exert a force that pushes the wire at right angles to the permanent magnetic field.

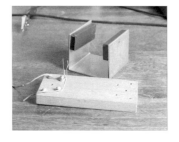

Making an electric motor

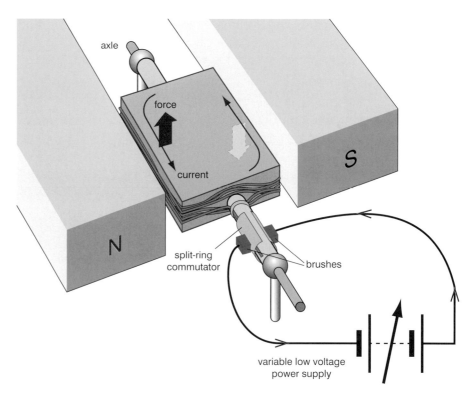

A motor coil as set up in the diagram will be forced round as indicated by the arrows (1 and 2 on page 151). The split-ring commutator ensures that the motor continues to spin. Without the commutator, the coil would rotate 90° and then stop. This would not make a very useful motor. The commutator reverses the direction of the current through the coil at just the right point (3) so that the forces on the coil flip around and continue the rotating motion (4).

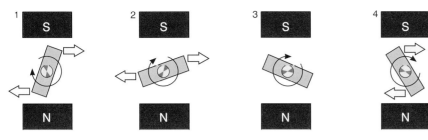

The motor can be used to show how the force on the current-carrying wires varies if you change the conditions. If you increase the current or if you increase the strength of the magnetic field, then the force is larger and the motor spins faster. If you change the direction of the current or if you reverse the magnetic field, the direction of the force is reversed and the motor spins in the opposite direction.

A* EXTRA

• Fleming's left-hand rule relates the movement of a coil to the direction of the permanent magnetic field and the direction of the current flowing in the coil.

REVIEW QUESTIONS

Q1 The diagram (right) shows a simple electromagnet made by a student. Suggest two ways in which the electromagnet can be made to pick up more nails.

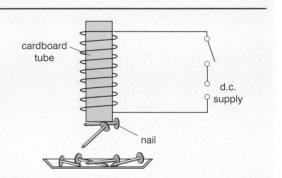

Q2 The diagram below shows an electric bell. Explain how the bell works when the switch is closed.

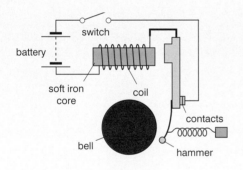

Q3 The diagram (right) shows a transformer.
  a What material is used for the transformer core?
  b What happens in the core when the primary coil is switched on?
  c What happens in the secondary coil when the primary coil is switched on?
  d If the primary coil has 12 turns and the secondary coil has 7 turns, what will the primary voltage be if the secondary voltage is 14 V?

Q4 Two students are using the equipment shown in the diagram (right). They cannot decide whether it is an electric motor or a generator. Explain how you would know which it is.

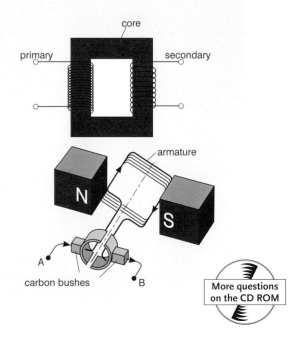

Examination questions are on page 156.

More questions on the CD ROM

6 CATHODE RAY OSCILLOSCOPES

Cathode rays

Cathode rays were discovered in the late 1800s. J.J. Thomson discovered that these rays consisted of a stream of **electrons** emitted from a heated cathode (a negative terminal).

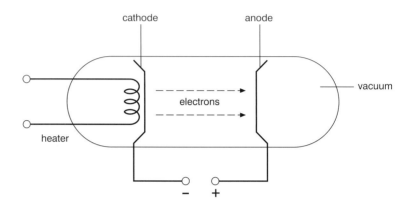

The positive terminal (the anode) attracts the electrons from the cathode. The cathode is heated to increase the average energy of the electrons in the cathode, which means that electrons will spontaneously jump out of the surface of the metal. The process of emitting electrons from a heated cathode is called **thermionic emission**.

If there is a hole in the anode, a beam of electrons shoots through. The whole arrangement is then called an **electron gun**.

An electron gun.

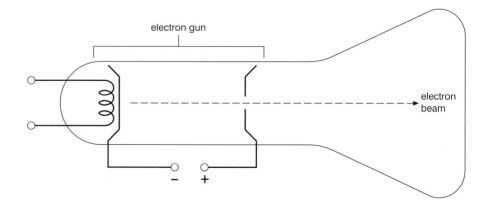

DEFLECTING THE BEAM

An electron beam is equivalent to an electric current, but without the wire. It can be deflected by **other electric charges** or by magnetic fields. In this diagram, the magnetic field is at right angles to the electron beam, so the beam is deflected. This is an example of the **motor effect** (see page 151). Check that it is obeying Fleming's left-hand motor rule, with First finger for Field, seCond finger for Current and thuMb for movement. Be careful over the question of which way the current is flowing.

- The electron beam is an electric current. We can link the *number* of electrons moving per second to the *current* of the beam. Using the formula $I = \frac{Q}{T}$, if there is a current of 1.0 A, then there must be 1 coulomb of electric charge moving per second. 1 coulomb of electric charge is equivalent to 6.25×10^{18} electrons.

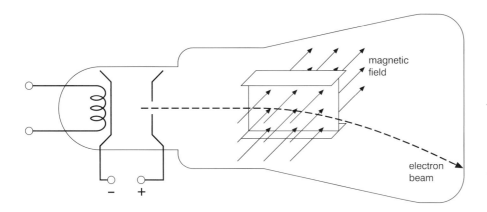

magnetic field

electron beam

Deflecting electrons with a magnetic field.

In this diagram, the metal plates are charged, attracting the electron beam towards the positive plate and repelling it away from the negative plate.

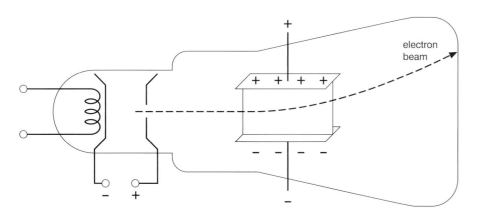

electron beam

Deflecting electrons with electric charges.

Simple treatment of the cathode ray oscilloscope

In a cathode ray oscilloscope, CRO, the electron beam is directed towards a **fluorescent screen**. Where the beam hits the screen, the coating on the screen absorbs the energy from the electrons and releases the energy as light – a dot appears on the screen.

The cathode ray oscilloscope.

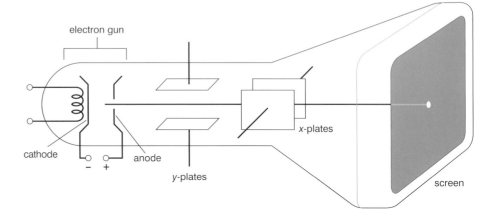

Two sets of metal plates are used to deflect the beam – one pair in the **vertical** direction (called the **y–plates**) and one pair in the **horizontal** direction (called the **x–plates**). By controlling the voltages on these sets of plates, the dot can be moved to any position on the screen.

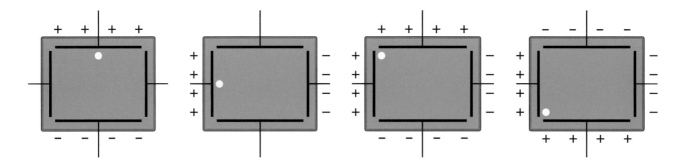

By changing the voltages on the X- and Y-plates, the dot can be moved around the screen.

A CRO has a circuit – the **time–base circuit** – that moves the dot across the screen, from left to right, at a constant speed and then returns the dot very quickly to the start. Repeating this process quickly means the dot appears as a **line** across the screen.

A CRO can be used as a very high resistance voltmeter. If an unknown voltage is connected to the y-plates, the deflection produced in the beam can be compared to reference measurements and a value for the voltage obtained.

With the time-base circuit switched on and a variable signal connected to the y-plates, a CRO can be used to show how a waveform varies. This can be used to measure the frequency of a signal.

REVIEW QUESTIONS

Q1 What is thermionic emission?

Q2 The diagram shows a CRO. Copy the diagram and add the labels in the correct places.
screen   x-plates   y-plates   anode   cathode   electron gun

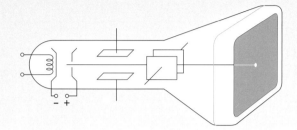

Q3 The diagram shows a trace on a CRO screen. The vertical setting is 0.2 V per division (square) and the timebase setting is 10 ms per division (square).
10 ms = 0.01 s.

a What is the amplitude of the signal in volts?

b What is the time period of the signal (the time for one complete cycle)?

c What is the frequency of the signal?

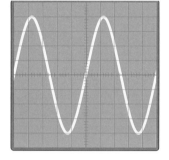

More questions on the CD ROM

Examination questions are on page 156.

EXAMINATION QUESTIONS

Q1 a Two non-conducting spheres, made of different materials, are initially uncharged. They are rubbed together. This causes one of the spheres to become positively charged and one negatively charged.

Describe, in terms of electron movement, why the spheres become charged.

_____

_____

_____ [2]

b Once charged, the two spheres are separated, as shown in Fig. 1.1.

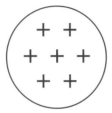 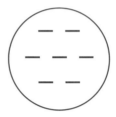

Fig. 1.1

On Fig. 1.1, draw the electric field between the two spheres. Indicate by arrows the direction of the electric field lines. [2]

c A conducting wire attached to a negatively charged metal object is connected to earth.

This allows 2.0×10^{10} electrons, each carrying a charge of 1.6×10^{-19} C, to flow to earth in 1.0×10^{-3} s.

Calculate

i the total charge that flows,

charge _____

ii the average current in the wire.

current _____ [3]

Q2 Fig. 2.1 shows a 12 V battery connected to a number of resistors.

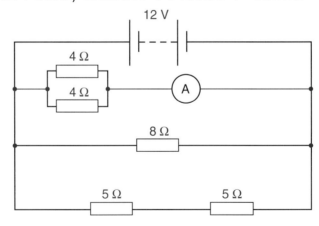

Fig. 2.1

a Calculate the current in the 8 Ω resistor.

current = _____ [2]

b Calculate, for the resistors connected in the circuit, the combined resistance of

i the two 5 Ω resistors,

resistance = _____

ii the two 4 Ω resistors.

resistance = _____ [2]

c   The total current in the two 4 Ω resistors is 6 A.
Calculate the total power dissipated in the two resistors.

power = _____   [2]

d   What will be the reading on a voltmeter connected across
  i   the two 4 Ω resistors,
      reading = _____
  ii   one 5 Ω resistor?
      reading = _____   [2]

e   The 8 Ω resistor is made from a length of resistance wire of uniform
cross-sectional area.
State the effect on the resistance of the wire of using
  i   the same length of the same material with a greater cross-sectional area,

  _____

  ii   a smaller length of the same material with the same cross-sectional area.

  _____   [2]

Q3 a i   What is the function of a transistor when placed in an electrical circuit?

  _____

  ii   Describe the action of a transistor.

  _____

  _____

  _____   [3]

  b i   In the space below, draw the symbol for an OR gate. Label the inputs and the
output.

  [1]

  ii   Describe the action of an OR gate that has two inputs.

  _____

  _____   [2]

Q4 a   Fig. 4.1 shows the screen of a CRO (cathode-ray oscilloscope).
The CRO is being used to display the output from a microphone.
The vertical scale on the screen is in volts.

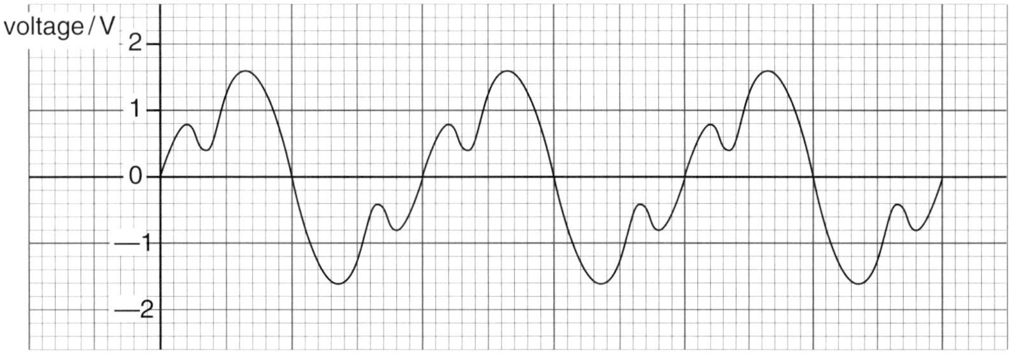

Fig. 4.1

  i   Describe the output from the microphone.

  _____

  _____

ii Use the graph to determine the peak voltage of the output.

_____

iii Describe how you could check that the voltage calibration on the screen is correct.

_____

_____ [4]

b Fig. 4.2 shows the screen of the CRO when it is being used to measure a small time interval between two voltage pulses.

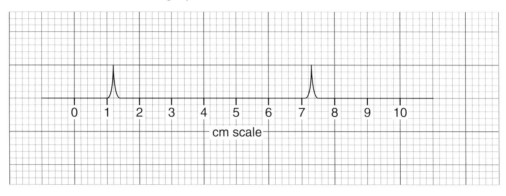

Fig. 4.2

i   What is the distance on the screen between the two voltage pulses?

_____

ii The time-base control of the CRO is set at 5.0 ms/cm.
    Calculate the time interval between the voltage pulses.
    time = _____

iii Suggest **one** example where a CRO can be used to measure a small time interval.

_____ [4]

Q5 Fig. 5.1 shows a 240 V a.c. mains circuit to which a number of appliances are connected and switched on.

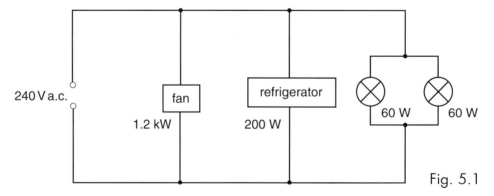

Fig. 5.1

a Calculate the power supplied to the circuit.
   power = _____ [1]

b The appliances are connected in parallel.
   i   Explain what connected *in parallel* means.

   _____

   _____

   ii State two advantages of connecting the appliances in parallel rather than in series.
       advantage 1 _____
       advantage 2 _____ [3]

c   Calculate
    i   the current in the refrigerator,
         current = _____
    ii   the energy used by the fan in 3 hours,
         energy = _____
    iii the resistance of the filament of one lamp.
         resistance = _____ [7]

Q6 Fig. 6.1 and Fig. 6.2 show two views of a vertical wire carrying a current up through a horizontal card. Points P and Q are marked on the card.

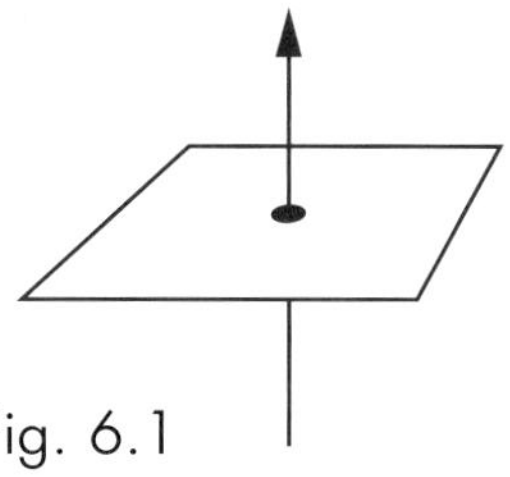

Fig. 6.1 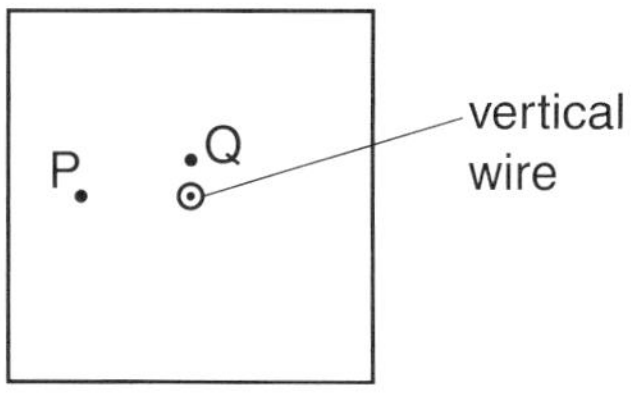   view from above the card   Fig. 6.2

a) On Fig. 6.2,
    i   draw a complete magnetic field line (line of force) through P and indicate its direction with an arrow,
    ii   draw an arrow through Q to indicate the direction in which a compass placed at Q would point. [3]

b   State the effect on the direction in which compass Q points of
    i   increasing the current in the wire,

    _____

    ii   reversing the direction of the current in the wire.

    _____ [2]

c   Fig. 6.3 shows the view from above of another vertical wire carrying a current up through a horizontal card. A cm grid is marked on the card. Point W is 1 cm vertically above the top surface of the card.

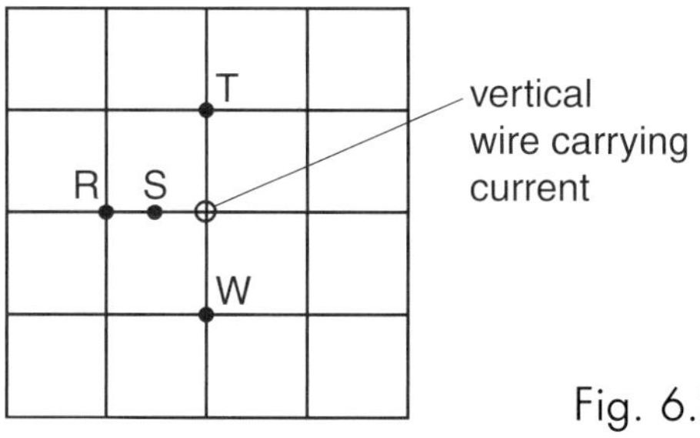

Fig. 6.3

State the magnetic field strength at S, T and W in terms of the magnetic field strength at R. Use one of the alternatives, **weaker**, **same strength** or **stronger** for each answer.
at S _____
at T _____
at W _____ [3]

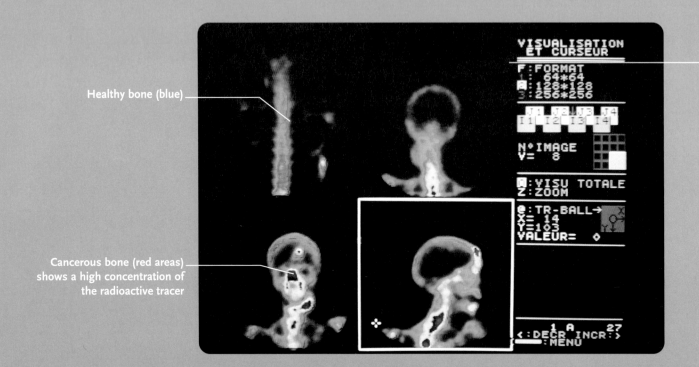

Healthy bone (blue)

Cancerous bone (red areas) shows a high concentration of the radioactive tracer

Seeing inside the body

Using radioactivity to 'see' inside people? Isn't radioactivity dangerous?

Strange as it may sound, using radioactivity, particularly gamma radiation, provides a powerful tool to help medical staff during diagnosis. Whilst it is true that gamma radiation can have harmful effects on living tissue, gamma imaging can provide information about what is happening inside a patient without the need for surgery. This non-invasive method is therefore quicker and overall has less risks than surgery. Of course, care is taken to minimise the radiation the patient receives, through careful monitoring of the dose and using a material with a suitable half-life

So how does it work? For an example, think about a bone scan. Bone tissue is good at absorbing phosphorus compounds, so the patient is injected with a radioactive material that contains phosphorus. This material will collect and form 'hot spots' at bone sites with high metabolic activity. High metabolic activity in bones might be connected to the growth of a tumour. These sites can be detected from outside the body using a gamma detector and so the medical staff have good information upon which to base further investigations.

ATOMIC PHYSICS

Gamma camera scans of the skull and spine of a
person suffering from multiple bone cancer.
This colour-coded image represents the position and
intensity of radiation emitted from a short-lived
radioactive tracer that concentrates in bone—more
strongly so in cancerous bone

1 RADIOACTIVITY

Videos & questions
on the CD ROM

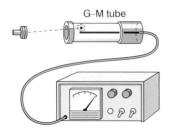

G–M tube

Radioactivity is measured using
a Geiger–Müller tube linked to
a counter.

Detection of radioactivity

All ionising radiation is invisible to the naked eye, but it affects
photographic plates. Individual particles of ionising radiation can
be detected using a Geiger–Müller tube.

There is *always* ionising radiation present. This is called **background
radiation**. Background radiation is caused by radioactivity in soil, rocks
and materials like concrete, radioactive gases in the atmosphere and
cosmic rays, which come from somewhere in outer space, though we are
still not sure *exactly* where.

Characteristics of the three kinds of emission

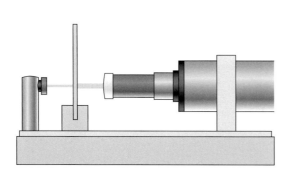

Radiation measurement device
with beta emitter, aluminium
sheet and meter.

Inside the atom the central **nucleus** of positively charged
protons and neutral **neutrons** is surrounded by shells, or
orbits, of **electrons**. Most nuclei are very stable, but some
'decay' and break apart into more stable nuclei. This
breaking apart is called **radioactive decay**. Atoms whose
nuclei do this are **radioactive**.

When a radioactive nucleus decays it may emit one or
more of the following:
* alpha (α) particles
* beta (β) particles
* gamma (γ) rays.

A stream of these rays is referred to as **ionising radiation** (often called
nuclear radiation, or just 'radiation' for short).

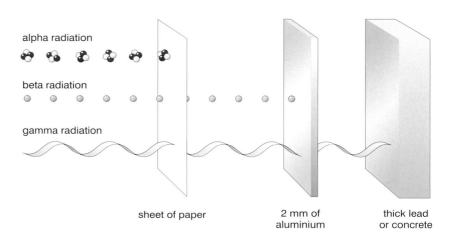

alpha radiation

beta radiation

gamma radiation

sheet of paper   2 mm of   thick lead
                        aluminium   or concrete

	alpha (α)	beta (β)	gamma (γ)
Description	A positively charged particle, identical to a helium nucleus (two protons and two neutrons)	A negatively charged particle, identical to an electron	Electromagnetic radiation. Uncharged
Penetration	4–10 cm of air. Stopped by a sheet of paper	About 1 m of air. Stopped by a few mm of aluminium	Almost no limit in air. Stopped by several cm of lead or several metres of concrete
Effect of electric and magnetic fields	Deflected*	Deflected* considerably	Unaffected – not deflected

* Note: Alpha particles and beta particles are deflected by magnetic and electric fields in the same manner that electrons are deflected in the cathode ray tube. They are travelling much faster so the deflections are smaller. Beta particles actually are electrons, so they are bent the same way. Alpha particles carry a positive charge, so they are bent the opposite way.

Radioactive decay

The **activity** of a radioactive source is the number of ionising particles it emits each second. Over time, fewer nuclei are left in the source to decay, so the activity drops. The time taken for half the radioactive atoms to decay is called the **half-life**. The activity of a radioactive source is measured in bequerels (Bq).

Half-life

Starting with a pure sample of radioactive nuclei, after one half-life half the nuclei will have decayed. The remaining undecayed nuclei still have the same chance of decaying as before, so after a second half-life half of the remaining nuclei will have decayed. After two half-lives a quarter of the nuclei will remain undecayed.

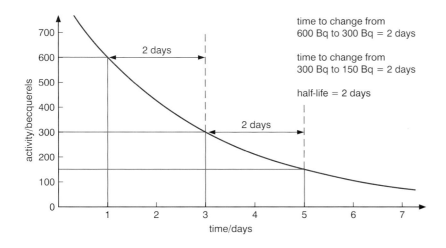

time to change from 600 Bq to 300 Bq = 2 days

time to change from 300 Bq to 150 Bq = 2 days

half-life = 2 days

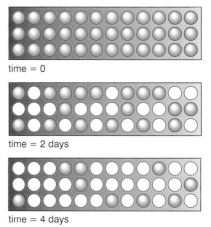

The half-life is 2 days. Half the number of radioactive nuclei decays in 2 days.

WORKED EXAMPLE

A radioactive element is detected by a Geiger–Müller tube and counter as having an activity of 400 counts per minute. Three hours later the count is 50 counts per minute. What is the half-life of the radioactive element?

Write down the activity and progressively halve it.
Each halving of the activity is one half-life:

0	400 counts
1 half-life	200 counts
2 half-lives	100 counts
3 half-lives	50 counts

3 hours therefore corresponds to 3 half-lives and 1 hour therefore corresponds to 1 half-life.

Safety precautions

Alpha, beta and gamma radiation can all damage living cells. Alpha particles, due to their strong ability to ionise other particles, are particularly dangerous to human tissue. Gamma radiation is dangerous because of its high penetrating power. However, cells have repair mechanisms that make ordinary levels of radiation relatively harmless.

Nevertheless, radiation can be very useful – it just needs to be used *safely*.

Safety precautions for handling radioactive materials include:
- Use forceps to hold radioactive sources – don't hold them directly.
- Do not point radioactive sources at living tissue.
- Store radioactive materials in lead-lined containers – and lock the containers away securely.
- Check the surrounding area for radiation levels above normal background levels.

High levels of radiation are extremely hazardous, and people handling highly radioactive materials must wear special film badges (containing photographic film) that monitor the dose that they are receiving. They may need to wear protective clothing, perhaps containing sheets of lead, and they will need to shower and check for radioactivity on their bodies at the end of each shift.

USES OF RADIOACTIVITY

Gamma rays can be used to kill bacteria. This is used in **sterilising** medical equipment and in preserving food. Due to the ability of gamma rays to penetrate material, the full thickness of the food can be treated even after it has been packaged.

A **smoke alarm** includes a small radioactive source that emits alpha radiation. The radiation produces ions in the air which conduct a small electric current. If a smoke particle absorbs the alpha particles, it reduces the number of ions in the air, and the current drops. This sets off the alarm. Alpha particles must be used because beta or gamma rays would pass through the air without producing enough ions.

Beta particles are used to monitor the **thickness** of paper or metal. The number of beta particles passing through the material is related to the thickness of the material. Alpha particles would not pass through the paper, and gamma rays would pass through almost unimpeded.

Nuclear power can generate enormous quantities of energy from the radioactivity inside elements, uranium in particular. There are considerable problems still to be overcome on waste disposal.

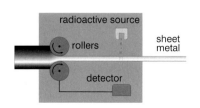

Sheet thickness control.

A gamma source is placed on one side of a **weld** and a photographic plate on the other side. Gamma rays pass through the metal, and weaknesses in the weld will show up on the photographic plate.

In **radiotherapy** high doses of radiation are fired at cancer cells to kill them. For cancers deep inside the body, gamma rays are used. Here, as in the case of X-rays, the radiation that can cause cancer is also an important tool in treating it.

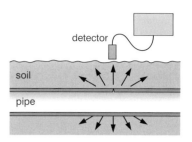

detector

soil

pipe

Tracers detect leaks.

In all of these above cases the radioactive material used must have a fairly long lifetime so that the device will function for long enough. For example, Americium-241, the material used in smoke detectors, has a half-life of 432 years, which is ample for the life of a smoke detector.

Tracers are radioactive substances with half-lives and radiation types that suit the job they are used for. The half-life must be long enough for the tracer to spread out and to be detected after use but not so long that it stays in the system and causes damage.
- **Medical tracers** are used to detect blockages in vital organs. A gamma camera is used to monitor the passage of the tracer through the body. Doctors often use Technetium-99m, with a half-life of 6 hours.
- **Agricultural tracers** monitor the flow of nutrients through a plant.
- **Industrial tracers** can measure the flow of liquid and gases through pipes to identify leakages.

RADIOACTIVE DATING

Igneous rock contains small quantities of uranium-238 – a type of uranium that decays with a half-life of 4500 million years, eventually forming lead. The ratio of lead to uranium in a rock sample can be used to calculate the age of the rock. For example, a piece of rock with equal numbers of uranium and lead atoms in it must be 4500 million years old – but this would be unlikely as the Earth itself is only 4500 million years old.

Carbon in living material contains a constant, small amount of the radioactive isotope **carbon–14**, which has a half-life of 5700 years. When the living material dies, the remaining carbon-14 atoms slowly decay. The ratio of carbon-14 atoms to the non-radioactive carbon-12 atoms can be used to calculate the age of the plant or animal material. This method is called **radioactive carbon dating**.

This woman's body was preserved in a bog in Denmark. Radioactive carbon-14 dating showed that she had been there for over 2000 years.

REVIEW QUESTIONS

Q1 What type of radiation is used in
a smoke detectors   b thickness measurement   c weld checking?

Q2 This question is about tracers.
a What is a tracer?
b The table below shows the half-life of some radioactive isotopes.
Using the information in the table only, state which one of the isotopes is most suitable to be used as a tracer in medicine. Give a reason for your choice.

Radioactive isotope	Half-life
lawrencium-257	8 seconds
sodium-24	15 hours
sulphur-35	87 days
carbon-14	5700 years

Q3 a When uranium-238 in a rock sample decays what element is eventually produced?
b Explain how the production of this new element enables the age of the rock sample to be determined.

More questions on the CD ROM

Examination questions are on page 170.

2 THE NUCLEAR ATOM

Atomic model

At the beginning of the 20th century, scientists knew that the atom contained positive and negative charges, but the structure was a great mystery. An experiment by Rutherford discovered what is now called Rutherford scattering. This effect was a surprising discovery, but one that greatly advanced our understanding of the atom.

(The actual work was done by two students of his, Geiger and Marsden. We were to hear more of Geiger later in his career.)

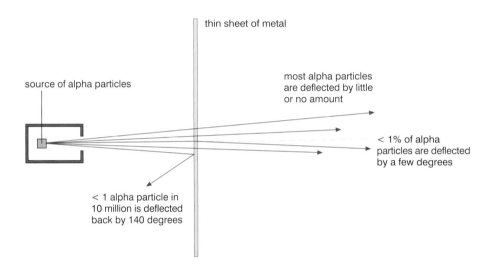

What they discovered was that almost all of the alpha particles got through the thin metal sheet with no difficulty, but that perhaps one particle in a million was scattered by a relatively massive object that sent it off at a wide angle, and perhaps even back the way it had come. This told them that the atom consists of a nucleus that contains almost all of the mass of the atom. Because so few alpha particles were scattered by the nucleus, it had to be extremely small, surrounded by a cloud of extremely light electrons.

We are left with the slightly disturbing thought that almost all of a solid object is actually empty space, loosely filled with electrons, with a tiny nucleus at the centre of each atom. If all of the space were taken out of an atom (and this can happen in a neutron star) then matter would have a density of 300 *million* tonnes per cubic centimetre.

The nucleus is made of protons and neutrons, bound together by an extremely strong force, far stronger than gravity, magnetism or electricity, and completely different from any of them.

There are several hundred different **nuclides** (types of nucleus), each with a different number of protons and neutrons. Nuclides tend to have rather more neutrons than protons. The majority of nuclides are radioactive. Most atoms on earth are not radioactive, of course, but these atoms are chosen from the limited number of nuclides that are stable.

Nucleus

You can write down nuclear changes as **nuclear equations**. Each nucleus is represented by its chemical symbol with two extra numbers written before it. Here is the symbol for Radium-226:

the top number is the **mass number** (the total number of protons and neutrons)

the bottom number is the **atomic number** (the number of protons)

$$^{226}_{88}\text{Ra}$$

The mass numbers and atomic numbers must balance on both sides of a nuclear equation.

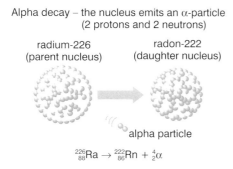

Alpha decay – the nucleus emits an α-particle (2 protons and 2 neutrons)

radium-226 (parent nucleus)   radon-222 (daughter nucleus)

alpha particle

$$^{226}_{88}\text{Ra} \rightarrow {}^{222}_{86}\text{Rn} + {}^4_2\alpha$$

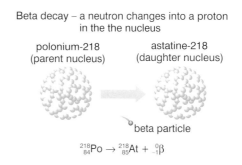

Beta decay – a neutron changes into a proton in the the nucleus

polonium-218 (parent nucleus)   astatine-218 (daughter nucleus)

beta particle

$$^{218}_{84}\text{Po} \rightarrow {}^{218}_{85}\text{At} + {}^0_{-1}\beta$$

Isotopes

There are two non-radioactive types of copper nuclei, $^{63}_{29}\text{Cu}$ and $^{65}_{29}\text{Cu}$. The first type contains 29 protons and 63 nucleons, hence (63 – 29) = 34 neutrons. The second type contains 29 protons and 36 neutrons. These two types of copper are known as isotopes of copper.

Isotopes contain the same number of protons but a different number of neutrons.

In naturally occurring copper, just under 70 per cent of the nuclei are $^{63}_{29}\text{Cu}$, and just over 30 per cent are $^{65}_{29}\text{Cu}$. Because they both contain 29 protons, they are surrounded by 29 electrons in the same pattern. It is the structure of the electrons around the nucleus that fix how the chemistry will work, so these two isotopes have the same chemistry, forming blue crystals of copper II sulfate, etc.

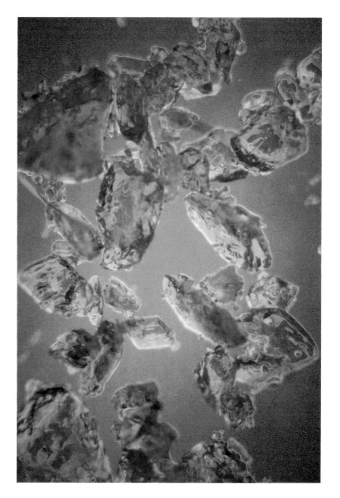

Tiny crystals of copper II sulfate under a microscope.

In addition there are various radioactive isotopes of copper. Because these do not have the ideal number of neutrons the nucleus is unstable. There are nine radioactive isotopes, with mass numbers that vary from $^{59}_{29}$Cu to $^{69}_{29}$Cu. These two extremes are very unstable with half-lives of a few minutes. $^{64}_{29}$Cu has a half-life of 12 hours. So all of the radioactive isotopes of copper are extremely radioactive. Some radioactive isotopes of other elements are much less radioactive, and have half-lives measured in years, if not thousands of years.

This copper statue contains 69 per cent $^{63}_{29}$Cu nuclei and 31 per cent $^{65}_{29}$Cu nuclei.

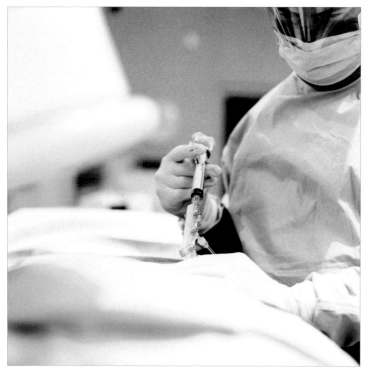

Radioactive iodine is injected in the treatment of cancer.

USES OF ISOTOPES

Radioisotopes have the same chemistry as non-radioactive isotopes of the same element. This can be very valuable in research as well as medicine. A famous example is the use of radioactive iodine in treatment of diseases of the thyroid.

The thyroid gland in the neck can grow dangerously large. Because the cells of the thyroid absorb far more iodine than other parts of the body, the thyroid can be targeted by injecting the body with radioactive iodine. Iodine-131 with a half-life of 8 days is used. This iodine is absorbed by some of the cells of the thyroid, and kills them as it decays and gives off beta particles.

There are many other uses of isotopes, such as:
– smoke detectors: they allow the smoke from a fire to be detected at an early stage and this early warning saves lives;
– food irradiation: a method of treating food in order to make it safer to eat and have a longer shelf life;
– agricultural applications: they are used to help understand chemical and biological processes in plants;
– archaeological dating of bones and artefacts.

REVIEW QUESTIONS

Q1 Copy and complete this table to show the particles in these atoms.

Atom	Symbol	Number of protons	Number of neutrons	Number of electrons
Hydrogen	^1_1H			
Carbon	$^{12}_6C$			
Calcium	$^{40}_{20}Ca$			
Uranium	$^{238}_{92}U$			

Q2 The graph shows how the activity of a sample of sodium-24 changes with time. Activity is measured in becquerels (Bq).
a Sodium-24 has an atomic number of 11 and a mass number of 24. What is the composition of the nucleus of a sodium-24 atom?
b Use the graph to work out the half-life of the sodium-24.

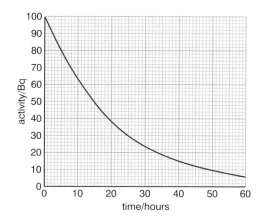

Q3 The following equation shows what happens when a nucleus of sodium-24 decays.

$$^{24}_{11}Na \rightarrow {}^x_y Mg + {}^0_{-1}\beta$$

a What type of nuclear radiation is produced?
b What are the numerical values of x and y?

Examination questions are on page 170.

More questions on the CD ROM

EXAMINATION QUESTIONS

Q1 a A radioactive isotope emits only α-particles.

    i In the space below, draw a labelled diagram of the apparatus you would use to prove that no β-particles or γ-radiation are emitted from the isotope.

    ii Describe the test you would carry out.

_____

_____

_____

_____

    iii Explain how your results would show that only α-particles are emitted.

_____

_____

_____ [6]

b Fig. 1.1 shows a stream of α-particles about to enter the space between the poles of a very strong magnet.

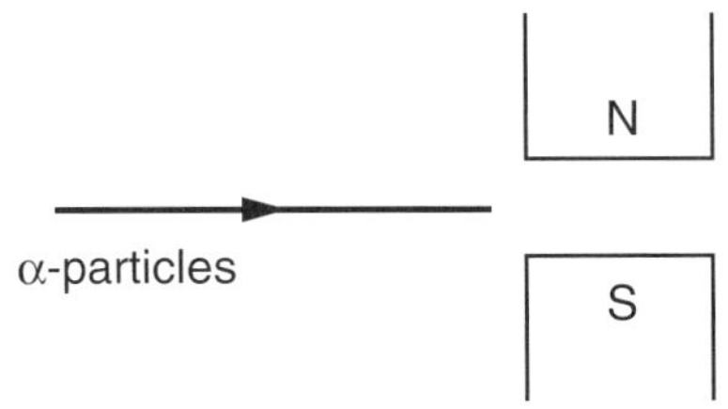

N

α-particles →

S

Fig. 1.1

Describe the path of the α-particles in the space between the magnetic poles.

_____

_____

_____ [3]

Q2 a A sodium nucleus decays by the emission of a β-particle to form magnesium.

    i Complete the decay equation below.

$$^{24}_{11}\text{Na} \rightarrow \text{Mg} +$$

    ii Fig. 2.1 shows β-particles from sodium nuclei moving into the space between the poles of a magnet.

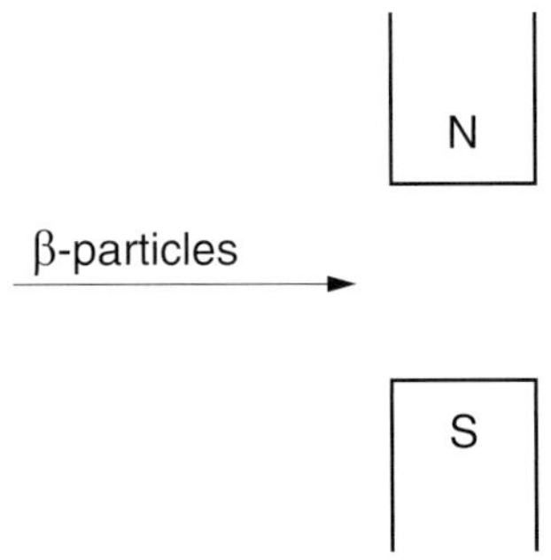

N

β-particles →

S

Fig. 2.1

Describe the path of the β-particles between the magnetic poles.

_____

_____ [5]

b Very small quantities of a radioactive isotope are used to check the circulation of blood by injecting the isotope into the bloodstream.
  i Describe how the results are obtained.

_____

_____

_____

_____

  ii Explain why a γ-emitting isotope is used for this purpose rather than one that emits either α-particles or β-particles.

_____

_____

_____ [4]

EXAM PRACTICE AND ANSWERS

EXAM PRACTICE AND ANSWERS

EXAM TIPS

Read each question carefully; this includes looking in detail at any **diagrams**, **graphs** or **tables**.
- Remember that any information you are given is there to help you.
- Underline or circle the **key words** in the question and **make sure you answer the question that is being asked.**

Make sure that you understand the meaning of the **'command words'** in the questions. For example:
- **'Describe'** is used when you have to give the main feature(s) of, for example, a process or structure;
- **'Explain'** is used when you have to give reasons, e.g. for some experimental results;
- **'Suggest'** is used when there may be more than one possible answer, or when you will not have learnt the answer but have to use the knowledge you do have to come up with a sensible one;
- **'Calculate'** means that you have to work out an answer in figures.

Look at **the number of marks** allocated to each question and also the **space provided**.
- Include at least as many points in your answer as there are marks. If you do need more space to answer, then use the nearest available space, e.g. at the bottom of the page, making sure you write down which question you are answering. **Beware of continually writing too much because it probably means you are not really answering the questions.**

Don't spend so long on some questions that you don't have time to finish the paper.
- You should spend approximately **one minute per mark**. If you are really stuck on a question, leave it, finish the rest of the paper and come back to it at the end.

In short answer questions, or multiple-choice type questions, **don't write more than you are asked for**.
- In some exams, examiners apply the rule that they only mark the first part of the answer written if there is too much. This means that the later part of the answer will not be looked at.
- In other exams you would not gain any marks if you have written something incorrect in the later part of your answer, even if the first part of your answer is correct. This just shows that you have not really understood the question or are guessing.

In calculations always show your working.
- Even if your final answer is incorrect you may still gain some marks if part of your attempt is correct. If you just write down the final answer and it is incorrect, you will get no marks at all.
- Also in calculations write your answer to as many **significant figures** as are used in the question.
- You may also lose marks if you do not use the correct **units**.

Aim to use **good English** and **scientific language** to make your answer as clear as possible.
- In short answer questions, just one or two words may be enough, but in longer answers take particular care with capital letters, commas and full stops.
- There should be an icon in the margin to warn you where there are separate marks for the quality of your English.
- If it helps you to answer clearly, do not be afraid to also use **diagrams** in your answers.

Some questions will be about scientific **ideas** and how scientists use **evidence**.
- In these questions you may be given some information about an unfamiliar situation.
- The answers to this type of question usually link to one of four areas: how scientists communicate their ideas; how scientific ideas can reflect the society in which the scientists work; how scientists can give different interpretations to the same evidence; how science can answer some questions but not others.

When you have finished your exam, **check through** to make sure you have answered all the questions.
- Cover your answers and read through the questions again and check your answers are as good as you can make them.

EXAM QUESTIONS AND STUDENTS' ANSWERS (CORE)

I a) Circle the names of two materials which are attracted to magnets.

aluminium ⟨brass⟩ ✗ copper ⟨iron⟩ ✓ steel tungsten [2]

b) The diagram shows a pattern of lines around a magnet.

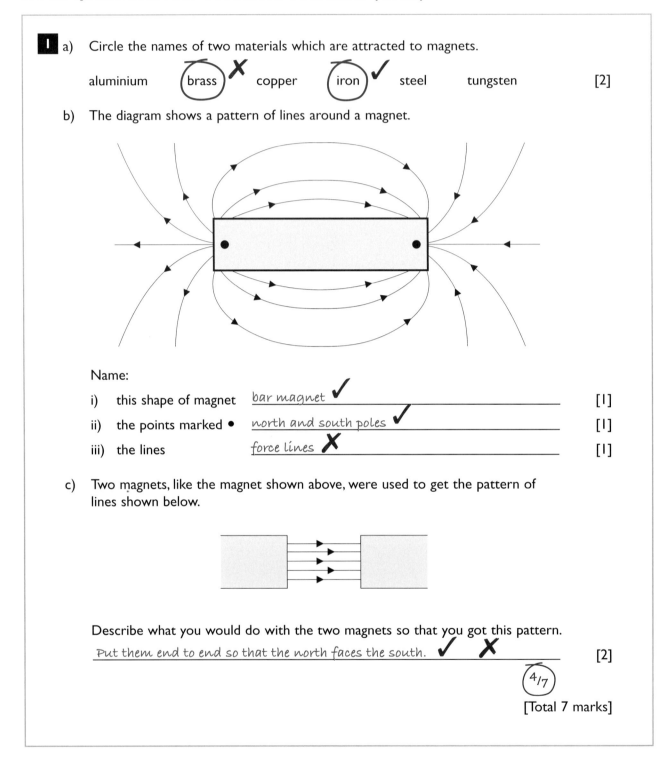

Name:

i) this shape of magnet   *bar magnet* ✓   [1]

ii) the points marked •   *north and south poles* ✓   [1]

iii) the lines   *force lines* ✗   [1]

c) Two magnets, like the magnet shown above, were used to get the pattern of lines shown below.

Describe what you would do with the two magnets so that you got this pattern.

Put them end to end so that the north faces the south. ✓ ✗   [2]

(4/7)

[Total 7 marks]

How to score full marks

a) One mark has been given for correctly choosing iron as a material that is attracted to magnets. Brass is incorrect – steel should have been the other choice from the list. There are only four magnetic metals – make sure you learn them carefully. **Also, the question asked for two names – make sure you do not give more than this.**

b) Two marks awarded. In part i) bar magnet is clearly acceptable, 'rectangular magnet' would also have scored the mark. In part ii), the key word is 'poles', the region of the magnet where the magnetism is strongest. The north and south given in the answer are correct and could even have been labelled on the diagram. Take care – writing 'north and south' on its own would *not* have scored the mark – 'north and south' are *directions* and can only score the mark here when the word 'poles' is added. In general, **make sure that you learn technical terms carefully and that you use the full**

term – this is where lots of practice questions can be very helpful. In part iii) no mark is awarded. The best answer would be 'magnetic field lines'. If the student had written '*magnetic* force lines' he or she would just have scored the mark, but as it stands it is wrong – there are many situations that might show 'force lines' without being linked to magnetism.

c) One mark is awarded. The student has correctly realised that the poles must be different for the magnetic field lines to form this pattern. However, the student has not made a second point about the situation – **always take notice of how many marks are available.** Possibilities for the second mark include: the north pole must be on the left (remember that field lines point from north to south) or that the magnets must be held apart (remember that the opposite poles would attract each other).

A mark of 4 out of a possible 7 corresponds to a grade F on this question.

QUESTIONS TO TRY (CORE)

1 a) In the box are the names of five waves.

| infrared | microwaves | ultrasonic | ultraviolet | **X-rays** |

Which wave is used to:

i) send information to a satellite?   [1]

ii) toast bread?   [1]

ii) clean a valuable ring?   [1]

b) The diagram shows four oscilloscope wave traces. The controls
of the oscilloscope were the same for each wave trace.

A 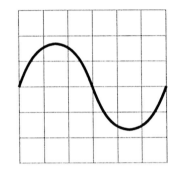   **B**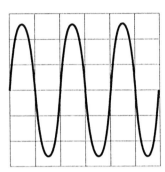

C 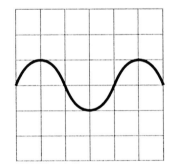   **D**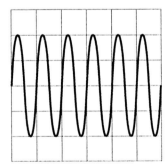

Which **one** of the wave traces, A, B C or D, has:

i) the largest amplitude?   [1]

ii) the lowest frequency?   [1]

c) The diagram shows a longitudinal wave in a stretched spring.

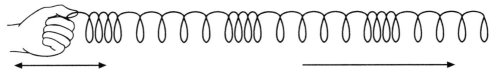

disturbance   direction of wave travel

Complete the sentence. You should put only **one** word in each space.

A longitudinal wave is one in which the .... causing the wave is

in the same .... as that in which the wave moves.   [2]

d) Which **one** of the following types of wave is longitudinal? Draw a ring around your answer.

light wave   **sound wave**   **water wave**   [1]

More questions
on the CD ROM

Answers are on page 188.

EXAM QUESTIONS AND STUDENTS' ANSWERS (CORE/EXTENDED)

I This question is about electrostatics.

a) There are two kinds of electric charge.

Write down the names of both types of electric charge.

<u>positive</u>   and   <u>negative</u> ✔   [1]

b) Leon wants to charge his plastic comb.

Write down one way he could charge his plastic comb.

<u>He could rub it.</u>   ✔ ✗

_____   [2]

c) Leon holds his charged comb near some small pieces of paper.

Look at the diagram.

Suggest what might happen to the paper.

<u>They stick to the comb.</u>   ✔

_____   [1]

d) Leon touches a metal radiator.

He gets an electric shock.

Describe how Leon gets an electric shock.

(One mark is for the correct use of scientific words.)

<u>The metal radiator is electric and gives Leon a shock.</u> ✗ ✗ ✗

[2+1]

e) Leon paints cars.

Static electricity is useful in spraying paint.

i) Write down **one other** use of static electricity.

<u>A photocopier</u> ✔ [1]

ii) Explain why static electricity is useful in spraying cars.

In your answer use ideas about electric charge.
(One mark is for linking ideas.)

<u>The paint is charged when it comes out of the sprayer. The car is also</u>
<u>charged with the opposite charge.</u> ✔ <u>This makes the paint stick to the</u>
<u>car much better.</u> ✔ [3+1]

6/12

[Total 12 marks]

How to score full marks

a) The correct response has been given. The symbols '+' and '−' would have been acceptable.

b) 'He could rub it' scores one mark, although 'by friction' would have been a stronger phrase to use. There is a second mark for saying that the comb should be rubbed against **an insulator** (or you could give an example of an insulator, such as cloth). **Always check the number of marks available.**

c) One mark has been awarded for the correct response. The student indicates correctly that there will be an **attraction** between the comb and the pieces of paper.

d) This is a very vague answer. There are three marks available, two for describing how Leon gets the electric shock. The correct response needs to realise that **Leon** has become **charged** (perhaps by friction against a carpet), that these charges **move** when he touches the radiator, **from Leon to the radiator.** The third mark is for using correct scientific words. Relevant words here are: charging, electrons, earth, earthing.

e) i) One mark has been awarded. Alternative correct responses would include inkjet printers, dust precipitators or crop spraying.

ii) Two marks have been awarded. The student seems to have an idea of what is happening, but has failed to use the correct scientific terms accurately. One mark has been awarded for the idea that opposite charges attract, but saying the paint 'sticks' to the car is not accurate enough to gain a second mark – the student needed to say that the paint is **attracted** to the car. In a similar way, saying the paint covers 'much better' is too vague. At this level, the student should refer to the paint being attracted to **the whole object**, even parts not in a direct line, or that **less paint is wasted.** Another approach would be to state that **like charges repel** (one mark) and so the paint forms **a fine spray** (one mark), which produces **an even coat** (one mark). The student does, however, gain the 'linking ideas' mark for linking the idea of opposite charges repelling to a consequence of this.

- A mark of 6 out of a possible 12 corresponds to a grade D on this Core/Extended question.

QUESTIONS TO TRY (CORE/EXTENDED)

I This question is about using scientific ideas and evidence.

The Octon Electricity Company wants to build a new power station.

It will burn fossil fuels.

Look at the map.

It shows the area around Octon where it wants to build the power station.

There are two available sites, **A** and **B**.

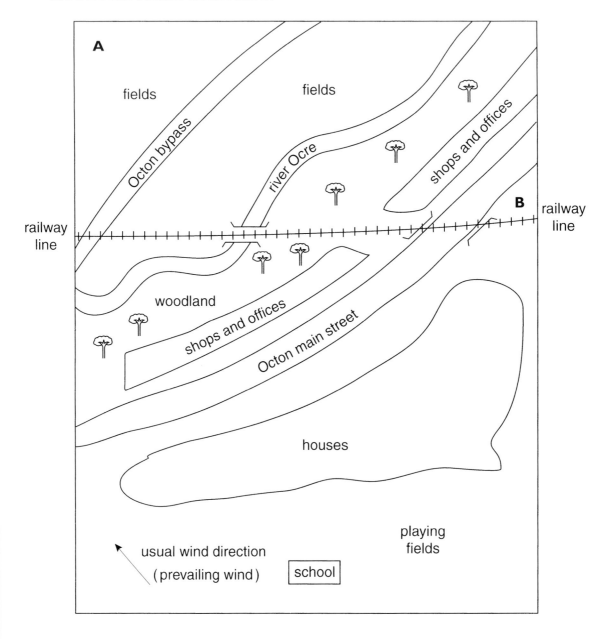

The company needs to identify the most suitable site. Suggest **two** questions that the company should ask itself and explain how its answers lead to the most suitable site. In your answer use scientific ideas. [4]

2

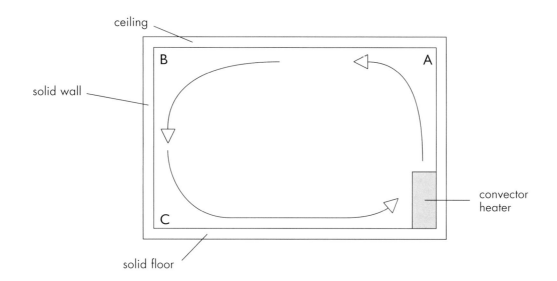

a) The room shown is heated by a convector heater. The arrows show the direction of movement of the air within the room.

Is the highest temperature at **A** or **B** or **C**?

_____   [1]

b) The room has a floor area of 20 m^2. The height of the room is 3 m and it contains air of mass 72 kg.

Calculate the density, in kilograms per cubic metre, of the air in the room.

_____

_____

_____   Density = _____ kg/m^3 [3]

c) The density of air changes with temperature.

Is the density of air lowest at **A** or **B** or **C**?   _____   [1]

d) Some heat energy within the room is transferred through the solid floor.

Name the energy transfer process taking place.

_____   [1]

More questions
on the CD ROM

Answers are on page 188–9.

EXAM QUESTIONS AND STUDENTS' ANSWERS (EXTENDED)

I Fig. I.1 shows a simple beam balance made from a pivot and a metre rule.

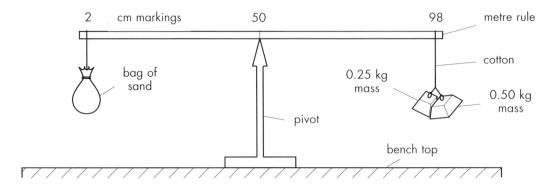

Fig. I.1

a) Find

   i) the mass of the bag of sand,

   mass = ___0.75 kg___ ✔

   ii) the weight of the bag of sand. (The acceleration of freefall is 10 m/s^2 .)

   weight = $m \times g = 0.75 \times 10 = 7.5$ N ✔✔   [3]

b) Explain, in terms of moments of forces, why the beam balances.

   <u>because the masses are the same</u> ✔

               ✗ ✗

     [3]

c) The cotton holding the 0.50 kg mass snaps and the mass falls to the bench. It strikes the bench at a speed of 1.2 m/s.

   Calculate its kinetic energy just before it hits the bench.

   $= \frac{1}{2} m v^2$

   $= 0.5 \times 0.5 \times 1.2^2$

   kinetic energy of the mass = $= 0.36$ ✔✔ ✗   [3]

d) On impact with the bench, the mass bounces up a small distance. Some transformation of energy occurs during the impact. State the forms of the energy just before and just after the impact.

   before: ___kinetic___ ✔

   after: ___sound___ ✔   ⑧/₁₁   [2]

   [Total II marks]

How to score full marks

a) Three marks awarded. In part i) the student has realised that the sand and the masses are the same distance from the pivot. This makes the calculations much easier. For this part of the question, it means that the mass on the left-hand side of the pivot must be the same as on the right-hand side. Take care to include all the masses on the right-hand side! **Remember to write down the unit** (kg in this case) if there is a space for it. In part ii) the student has used the correct equation and has calculated correctly, so two marks are given. The student has again given the correct unit.

b) One mark awarded – clearly this answer is too short. **Always check how many marks are available.** In this case three marks are available, so three points will be needed in the answer. **The number of lines given for the answer is usually a good guide to the length of response required.** The student does gain a mark (just) for saying the masses are the same – it would have been better to specify 'the masses hanging on either side of the pivot are the same'. The

student does not, however, go on to state that the masses are equal distances from the pivot (second mark) and so the turning effect, the moment, is the same on both sides (third mark).

c) Two marks awarded. The student uses the correct equation and calculates the value correctly. However, this time the student has not included the unit (joules, J), even though a space is provided, and so loses a mark. Remember that **almost all quantities in physics have a unit associated with them** – make sure you are familiar with them.

d) Two marks awarded. The question states that the mass has kinetic energy before the collision. After the collision, several possibilities exist. Some energy will be transferred as heat, the mass has bounced so it will still have some kinetic energy, it has bounced upwards so it has some gravitational potential energy. Since the question does not specify how many examples are required, it is probably best to keep the answer simple – **two marks are available, so give two answers.**

QUESTIONS TO TRY (EXTENDED)

1 Bobby goes on holiday in a caravan.

He uses a wind turbine to make electricity.

a) The turbine turns the simple a.c. generator shown below.

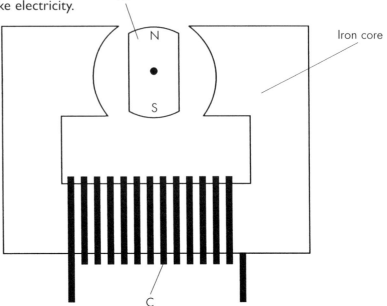

i) The term a.c. stands for **alternating current**.
Explain the difference between alternating current and direct current. [1]

_____

ii) Explain how the simple generator shown above generates alternating current.
Include the names of the parts labelled **M** and **C**. [4]

_____

_____

b) The generator is connected to the caravan by long cables.
Bobby is worried about energy losses in the cable.

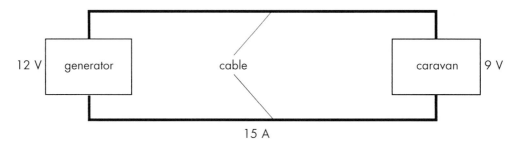

The generator produces a.c. at 12 V.

Bobby finds that when the current in the cable is 15 A, the voltage at the caravan is only 9 V.

i) State the equation which links voltage, current and resistance. [1]

_____

ii) Calculate the resistance of the **cable**. [2]

_____

2 a) All living things contain carbon atoms. All materials such as leather or wood, which come from living things, also contain carbon atoms.
Of all these carbon atoms, a tiny proportion is carbon-14.

The nuclear equation for the radioactive decay of carbon-14 is

$$^{14}_{6}C \rightarrow ^{14}_{7}N + ^{0}_{-1}e$$

Beta particles are emitted in this decay. How can you tell this from the equation?

_____

_____

_____ [2]

b) There are three forms of carbon: carbon-12, carbon-13 and carbon-14.
Complete the sentence.

These three forms are _____ of carbon. [1]

c) Radium-226 is a radioactive metal which decays by alpha emission to radon-222 which is a radioactive gas.

Complete the nuclear equation for this decay.

$$^{226}_{88}Ra \rightarrow \underline{\hspace{1cm}} Rn + \underline{\hspace{1cm}} He$$ [2]

3 A hot air balloon is tied to the ground by two ropes.
The diagram shows the forces acting on the balloon.
The ropes are untied and the balloon starts to move upwards.

a) Calculate the size of the unbalanced force acting on the balloon.
State the direction of this force. [2]

b) The mass of the balloon is 765 kg.
Calculate the initial acceleration of the balloon. [3]

c) Explain how the acceleration of the balloon changes during the first ten seconds of the flight. [3 + 1]

d) When the balloon is still accelerating, the balloonist throws some bags of sand over the side.
Explain how this affects the acceleration of the balloon. [2]

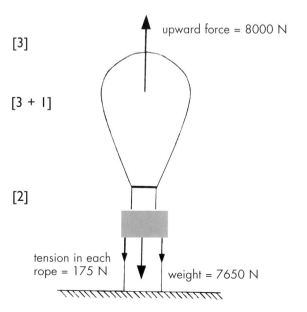

upward force = 8000 N

tension in each rope = 175 N

weight = 7650 N

More questions on the CD ROM

Answers are on page 189.

ANSWERS AND SOLUTIONS TO QUESTIONS TO TRY

Core (pages 178–9)

Q1 a (i) microwaves (1).

(ii) infrared (1).

(iii) ultrasonic (1).

Choosing the correct word from a list is very common at this level of question. Notice that all the waves given are part of the electromagnetic spectrum **except** for ultrasonic.

b (i) B (1).

(ii) A (1).

The wave with the largest amplitude is also the loudest and has the tallest waveform. The wave with the lowest frequency also has the lowest pitch and has the fewest waveforms on the screen.

c disturbance / vibration / movement (1). direction (1).

Describing the motion in longitudinal waves and transverse waves is tricky, so the examiners have tried to help by giving a sentence to complete and a diagram with some key words. **Always look carefully at any diagrams given**. Also remember that in a transverse wave the disturbance causing the wave is at right angles to the direction in which the wave moves.

d sound wave (1).

You **must** remember that sound is an example of a longitudinal wave. Almost all other waves studied at IGCSE are transverse, including all electromagnetic waves.

Core/Extended (pages 182–3)

Q1 There is a maximum of 4 marks: 2 marks maximum for relevant questions with further marks given for relevant reasons.

Any **two questions** from:

Question relating to environment

Question relating to transport

Question relating to resource availability, e.g. fuel / water / people / space

Relevant reasons linked to the questions asked:

Effect on environment – noise / pollution / smoke / named pollutants from a power station / cables and power lines

Relating to transport – availability of road or rail links / increase in pollutants

Relating to resources – use of fossil fuels rather than renewable resources (possibly giving examples) / availability of water for cooling / availability of people as staff for the plant.

The key is to give replies that are **sensible** and to relate them to particular pieces of scientific knowledge. It is easy to write answers that are too vague to score marks, such as 'Will it be good for Octon?', 'Will it be dangerous?'. Remember that this is a question asking about the wider implications of physics, so use your physics knowledge to guide you. Also, **study the diagram carefully**, there are clues to help you.

Q2 a A

The air is heated by the convector heater and the hot air will then rise. From the choices available, A will have the highest temperature. The air will be losing heat, and therefore have a lower temperature, as it is moved across to B. The air will only sink to C if it loses more heat and so has an even lower temperature.

b 1.2 (kg/m^3)   (3)

$$\text{Density} = \frac{\text{mass}}{\text{volume}}$$

or volume = area × height   (1)

$$\text{Density} = \frac{72}{60}$$ (1)

Here is a calculation that is worth 3 marks. To get the final answer, you need to calculate the volume of the room (area × height) and then use this number to calculate the density $\left(\frac{\text{mass}}{\text{volume}}\right)$.

It is always a good idea to write down your working out. It is very difficult to do calculations and transfer numbers in your head and, if you try, you increase your chances of making a mistake. Also, if you do get the final answer wrong, you may still get some marks for using the correct equations.

c A (1)

The air has the highest temperature at A, so it will have expanded most at that point. If it has expanded the most, then its density will be lowest.

d Conduction (1)

Candidates often get confused when using the terms conduction, convection and radiation. As a general, rough guide – through solids it will be conduction (the particles cannot move), through liquids and gases it will be convection (the particles are mobile). Heat radiation is emitted by any object that is warmer than its surroundings.

Extended (pages 186–7)

Q1 a (i) Alternating current keeps changing direction but direct current flows in a constant direction in a circuit. (1)

This is often answered very poorly, if at all. Common mistakes are to say that 'd.c. (direct current) is a steady current while a.c. (alternating current) is a changing current' – **this is wrong**. Another common mistake is to give the sources of each (e.g. d.c. from a battery, a.c. from the mains) rather than describing the **difference between them**.

(ii) Magnet (M) rotates (1), which changes the magnetic field / magnetism / field lines (1) of the iron core (1) and this causes / induces a current in the coil (C) (1).

There are several places in the specification where **you need to learn a sequence of events in order**. This is one instance. If you learn these parts carefully, you will avoid the common confusions that appear in candidates' answers – it is common to see phrases such as 'magnetic current' and 'the magnet is attracted to the electricity'.

b (i) Voltage = current × resistance (1)

(ii) $V = 12 - 9 = 3$ V (1)

$R = \dfrac{3}{15} = 0.2\ \Omega$ (1)

This question specifically asked for the equation, but **make sure you always write the equation anyway**. Here, the tricky bit is to realise that the voltage needed is 3 V. However, **if you write down your working** you will still receive the final mark even if you choose the wrong value for V.

Q2 a Any two from (1 mark each):

• They are electrons

• They are negative/negatively charged

• They have (almost) zero mass

This is recall using the clues in the nuclear equation given. The symbol given is for an electron, the '0' tells you it has almost no mass, the '-1' tells you it is negatively charged. Learn the symbols for alpha, beta and gamma radiation.

b Isotopes

This is basic recall. Spend some time learning basic facts and definitions.

c $^{226}_{86}$Rn (1)

$^{4}_{2}$He (1)

You should recognise the symbol for an alpha particle using He. You must learn the numbers that go with it, as mentioned in a) above. Having got these numbers in, you can calculate the numbers for Rn – remember that all the particles must end up somewhere. At the start of the equation there were 88 protons and 226 nucleons, so the totals on the right-hand side of the equation must add up to this.

Q3 a force = 350 N (1) upwards (1).

Total upwards force = 8000 N. When the ropes are untied, the tension in each falls to zero and so the downward force is just the weight of the balloon, 7650 N. The unbalanced force is therefore (8000 – 7650) N = 350 N. Do not forget to give the direction of this overall force.

b acceleration = force / mass (1)
= 350 N / 765 kg (1)
= 0.46 m/s^2 (1)

Recall $F = ma$ and re-arrange to get $a = \ldots$ The word 'initial' is part of the question because of what follows in part (c). Do not forget the unit of acceleration.

c The resistive force increases (1) reducing the size of the unbalanced force (1), so acceleration decreases (1). There is one communication mark for accurate spelling, punctuation and grammar (1).

The resistive force is air resistance, or drag. As the balloon moves upwards it has to push air out of the way. This creates a force pushing against the balloon. The faster the balloon moves, the more quickly it has to push air out of the way, so the greater the resistive force becomes. The overall (net) force becomes smaller as it is equal to (weight of balloon – resistive force) and so the acceleration reduces.

d An explanation to include: mass / downwards force decreases (1), causing an increase in acceleration (1).

Throwing the bags overboard reduces the weight of the balloon – it becomes lighter. The upward force does not change (because that is the lift provided by the air in the balloon), so the overall forces become more unbalanced – there is a larger overall upwards force, causing a larger acceleration. Remember to include the comment about the acceleration.

Answers and Solutions

GENERAL PHYSICS

1 Length and time (page 10)

Q1 The thinner edge means that the distance between the objects to be measured and the scale is reduced. It is therefore not so critical to get the eye vertically over the point being measured.

The ruler is thicker in the middle to make it stronger; to keep it straight; perhaps to make it easier to hold.

Q2 **a** 18 cm

b 0.055 or 0.056 cm^3

c 5 cm^2

d The inside diameter is 1.26 cm (or 1.3 cm).

Q3 **a** She said that the pendulum had swung one time when it had not swung at all. Hence she said 'Two' when it had swung once, etc.

b 1.33 s (or 1.3 s)

c None

2 Speed, velocity and acceleration (page 16)

Q1 **a** 600 km/h

b 500 km/h

c 150 km

Q2 **a** The speed remained constant.

If the line on a distance–time graph has a constant gradient then the speed is constant.

b 600 m.

This can be read off the graph: after 20 s the car had travelled 200 m, after 60 s it had travelled 800 m. So the distance travelled is
800 − 200 = 600 m.

c 15 m/s.

Speed = distance/time, v = 600/40 = 15 m/s. Don't forget the units.

d It stopped.

The line is horizontal during this time, indicating that the car was not moving (no distance was travelled).

Q3 **a** 2.0 m/s^2.

Acceleration = change in velocity/time,
a = (1.5 − 0)/0.75 = 2 m/s^2.

b 2.25 m.

Total distance = area under the line =
(1/2 × 0.75 × 1.5) + (1/2 × 2.25 × 1.5) = 2.25 m.

Note: the tractor showed constant acceleration from A to B and then constant deceleration from B to C.

3 Mass and weight (page 19)

Q1 **a** 600 N (or 588 N)

b 60 kg

c 228 N

d 60 kg

e 228 N

Q2 **a** Stay similar

b Increase

c Increase

d Increase

e Probably stay similar, though there is a case for decrease if, for example, jumping along turned out to be faster than running.

4 Density (page 23)

Q1 **a** Depends on the sample

b Float

c Depends on the sample

d Float

e Sink

f Sink

g Float

Q2 1900 kg/m^3

Q3 If the crown is pure gold the new water level will be 900 cm^3.

If the jeweller has cheated, the water level will be higher than 900 cm^3.

5 Forces (page 35)

Q1   **a**   Stage 1 – The skydiver is accelerating. The downward force of gravity is greater than the upward force caused by air resistance. Stage 2 – The skydiver is travelling at constant speed. The forces of gravity and air resistance are balanced. Stage 3 – The skydiver is slowing down. The force caused by the air in the parachute is greater than the force of gravity. Stage 4 – The skydiver is travelling at a constant speed. The forces are balanced again.

In questions of this sort the first thing to do is to decide which forces will be acting on the object. The next thing is to decide whether the forces are balanced or unbalanced. If they are unbalanced the skydiver will be either accelerating or decelerating. If they are balanced the skydiver will either be travelling at constant speed or not moving. In stage 2 the skydiver will have reached the terminal speed and because the forces are balanced will not accelerate or decelerate.

b   The force of gravity is balanced by the upward force of the ground on the sky diver.

The upward force from the ground is equal to the skydiver's weight.

Q2   **a**   The graph is a straight line from the origin to the 12 cm, 6.0 N point. After that it becomes increasingly curved and less steep. Don't be fooled into drawing a single 'best fit' line as there are two distinct parts to this graph.

b   The limit of proportionality should be marked on the graph at the end of the region that is a straight line. Proportional behaviour occurs all along the region where the graph is a straight line. Plastic behaviour occurs thereafter, though not necessarily immediately so.

c   If the spring returns to its original length then the behaviour was purely elastic. If it is longer, then the behaviour was partially elastic and partially plastic.

d   0.5 N/cm.

Q3   **a** 2.5 m/s^2.

Acceleration = change in speed/time, $a = (30 - 0)/12 = 2.5 \text{ m/s}^2$. Don't forget the units.

b   2500 N.

Force = mass x acceleration, $F = 1000 \times 2.5 = 2500$ N.

Q4   **a**   5000 N, approximately Northeast.

b   0.1 m/s.

Q5   800 Nm.

Moment = force x distance.

Q6   C.

C has the widest base and is heavier at the bottom (more glass).

6 Energy, work and power (page 43)

Q1   **a**   A source that cannot be regenerated – it takes millions of years to form.

A common mistake is to say that it is a source that 'cannot be used again'. Many energy sources cannot be used again but they can be regenerated (e.g. wood).

b   Coal, oil and natural gas.

These are the fossil fuels. Substances obtained from fossil fuels such as petrol and diesel are not strictly speaking fossil fuels.

c   Coal.

Coal is becoming increasingly more difficult to mine as more inaccessible coal seams are tackled. The 300 year estimate could be very optimistic.

Q2   **a**   The total energy used over this period has increased. There was a reduction in energy used from 1980 to 1983, but since then the consumption has increased.

b   The use of oil reduced up until about 1985, but has remained fairly constant since then. The use of coal has reduced as the use of natural gas and nuclear power has increased. The use of hydro power has remained fairly constant. The use of renewables was negligible in 1980, has increased but is still a small percentage of total energy consumption.

c   I would expect energy consumption to continue to increase. I would expect the use of renewable energy sources to increase, the use of coal to decrease.

Q3 **a** Any two from: strength of the wind, high ground, constant supply of wind, open ground.

Higher ground tends to be more windy than lower ground. However, the wind farm cannot be built too far away from the centres of population otherwise extra costs will be incurred.

b Advantage: renewable energy source, no air pollution. Disadvantage: unsightly, takes up too much space.

Wind turbines can be very efficient. However, they need to be reasonably small and so a large number are needed to generate significant amounts of electricity. Environmentally, although they produce no air pollution they do take up a lot of land.

Q4 12.5 m.

Work done = $F\,s$
$F = 400 \times 10 = 4000$ N
$s = W/F = 50\,000/4000 = 12.5$ m

Q5 **a** 6000 J.

Energy transferred = $30 \times 200 = 6000$ J.

b 33.3 W.

Power = work done/time taken = $6000/180 = 33.3$ W. Remember that a watt is 1 joule/s so the time must be in seconds.

Q6 **a** 10 500 J.

p.e. = $m\,g\,h = 35 \times 10 \times 30 = 10\,500$ J

b 24.5 m/s.

Assuming all the potential energy is transferred to kinetic energy:
k.e. = $1/2\,mv^2$, so $v^2 = 2$ k.e./m
$= 2 \times 10\,500/35 = 600$, and $v = 24.5$ m/s

c Some of the gravitational potential energy will have been transferred to thermal energy due to the friction between the sledge and the snow.

Energy must be conserved but friction is a very common cause of energy being wasted, that is, being transferred into less useful forms.

7 Pressure (page 49)

Q1 ordinary shoe heel = 160 kPa; elephant = 159 kPa; high-heeled shoe = 8000 kPa. The high-heeled shoe will damage the floor.

Q2 He is 15 m deep. If in sea water, the submarine would be slightly less than 15 m deep.

Q3 The increase in pressure is 29.4 kPa. The total pressure is 189.4 kPa.

THERMAL PHYSICS

1 Simple kinetic molecular model of matter (page 61)

Q1 **a** Solids keep their shape because their particles are closely packed together and are held in a rigid shape by the bonds between the particles, whereas the particles in liquids and gases are able to move.

b Gases fill their container because their particles move about freely, whereas the particles in a solid vibrate within their structure and those in a liquid only slide over each other.

Q2 **a** Evaporation of molecules with high k.e. lowers the average k.e. of the remaining molecules. A wet cloth provides a large area for evaporation.

b The draught removes the evaporated molecules and makes it less likely that they will re-enter the liquid.

c The air molecules have weight. This is usually obscured by the fact that they are floating in the atmosphere.

d If the can heats up, the pressure inside will increase until there is perhaps an explosion. The explosion will throw pieces of the fire at the person who disposed of the can.

Q3 **a** 800 litres.

b 1000 s or 16.7 minutes.

2 Thermal properties (page 75–6)

Q1 **a** The ruler will be the wrong length at other temperatures. For example, if it is hotter, then objects will appear to be shorter than they actually are.

b Different parts of the glass will expand by different amounts. This will stress the glass, which may crack.

c Water vapour molecules in the atmosphere will stick to the outside of the glass, and if they build up you will see the water.

d Even in the freezer, the ice is slowly evaporating. If the food is uncovered the water vapour will escape and the food will dry out.

Q2 a The end of the strip will move in one direction such that the steel strip is on the inside of the curve of the strip; the end will move the other way; the strip will return to its original position.

b Attach a suitable scale next to the end of the strip. Commercial thermometers use a very long strip bent into a spiral. The outside end of the spiral is attached to the instrument, and the inside end of the spiral is fixed to a pointer. The pointer rotates as the temperature changes.

c Place a switch contact near to the moving end of the bimetallic strip, on the aluminium side of the strip. Make the strip itself part of the switch by connecting one wire to the strip and the other wire to the contact. Alternatively, arrange for a manufactured switch to be mechanically operated by the end of the strip.

d The light would continuously switch on and off. Lamps of this type are available for flashlights.

Q3 a 18 °C; 30240 J.

b 60 g

c 32 °C; 8064 J

d 22176 J

e 369.6 J/g (or 370 J/g).

3 Transfer of thermal energy (page 81)

Q1 The thin layers will trap air between them. Air is an insulator and so will reduce thermal transfer from the body.

Remember that air is a very good insulator. Trapping it between layers of clothing means that convection is inhibited as well as conduction.

Q2 a Dish A shows least evaporation, dish B shows more evaporation, and dish C shows most evaporation.

b In dish A, some of the molecules in the water that have high k.e. manage to escape from the surface of the water. In dish B the dish absorbs infrared radiation from the Sun and warms the water. When the temperature of the water increases, the number of water molecules with high k.e. increases and more of them evaporate. In dish C, the just-evaporated water molecules are swept away by the wind and are less likely to return into the water.

c While the water is evaporating, much of the energy absorbed by the dish is carried away by the water molecules with high k.e. When the water has evaporated, this route for the heat energy to escape is blocked and so the heat energy is used to raise the temperature of the dish.

Q3 a Energy from the hot water is transferred from the inner to the outer surface of the metal by conduction. Energy is transferred through the still air by radiation.

In a question like this it is important to take the energy transfer stage by stage. Thermal transfer through a solid involves conduction. Thermal transfer through still air must involve radiation. In fact, convection would be occurring and it is very likely that the air behind the radiator would be moving (convention currents).

PROPERTIES OF WAVES, INCLUDING LIGHT AND SOUND

1 General wave properties (page 90)

Q1 a D.

The crest is the top of the wave.

b A and E.

The wavelength is the length of the repeating pattern.

c B.

The amplitude is half the distance between the crest and the trough.

d 512.

Frequency is measured in hertz and 1 Hz = 1 cycle (or wave) per second.

Q2 0.33 m.

$v = f \times \lambda$ or $\lambda = v/f$
$\lambda = 3 \times 10^8 / 9 \times 10^8 = 0.33$ m

Q3 **a** The water gets shallower 200 m ahead.

b The current is going right to left. The waves make the piece of wood bob up and down, but they do not move it along.

c If this is so, mobile phone signals cannot penetrate the walls of your house and must come in through the windows. Because mobile phones use short wavelengths, the waves cannot easily bend around buildings. Hence unless your window is facing a mobile phone transmitter the signal cannot get to your window.

2 Light (page 99)

Q1 **a** Light.

The sensitive cells form a part of the eye called the retina.

b Ultraviolet.

Suntan lotions contain chemicals that absorb some of the UV radiation from the Sun before it can act on the skin.

c Microwaves.

The key word here was 'rapid'. Electric cookers make use of infrared radiation for cooking but microwaves produce much more rapid cooking.

d X-rays.

A gamma camera would not be as good for this purpose as the gamma rays penetrate the bone as well as the flesh.

e Infrared.

Remote car locking sometimes uses infrared beams.

Q2 **a** In reflection light changes direction when it bounces off a surface. In refraction the light changes direction when it passes from one medium to another.

In reflection the angle of incidence and the angle of reflection are the same. In refraction the angle of incidence does not equal the angle of refraction, as the ray of light bends towards or away from the normal.

b (i)

(ii) The angle of incidence.

(iii) The normal line.

The ray of light will only be bent if it hits the block at an angle. In both cases the speed of the light in the block will be slower than in air.

Q3 **a**

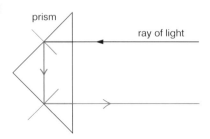

You should draw a normal line to ensure that for each reflection the angles of incidence and reflection are the same.

b Binoculars, bicycle reflector, cat's eyes.

Periscopes are also often made using prisms rather than mirrors.

c R, T, T.

Total internal reflection occurs only when the angle of incidence is equal to or greater than the critical angle.

Q4 **a** 2.07

b It will be slightly less than 21.7°.

c 28.9°

3 Sound (page 103)

Q1 **a** (i) Vibrations. (ii) Compressions and rarefactions travel through the air.

These are both common questions, so don't forget. Sound is a longitudinal wave. The compressions and rarefactions are small differences in air pressure.

b Sound waves cannot travel through a vacuum. Radio waves don't require a medium.

This is a big difference between sound waves and electromagnetic waves.

c The sound can bend round the buildings and other obstacles between you. In addition, air absorbs high frequencies quite strongly as they travel through it.

Q2 341 m/s (or 340 m/s)

Q3 **a** 1.3 m (or 1.33 m)

b 85 m

The glass block diagram (Q2 b (i)):

glass block

Ray 1   Ray 2

a

ELECTRICITY AND MAGNETISM

1 Simple phenomena of magnetism (page 111)

Q1 A magnetically hard material stays magnetic once it has been magnetised (e.g. steel). A magnetically soft material does not stay magnetic when the source of magnetisation is removed (e.g. pure iron).

Q2 The piece of metal could just be made of a soft magnetic material. This will be attracted if it is brought near to a magnet. Ranjit could turn the piece of metal around and see if it is still attracted (in which case it is not a magnet) or repelled (in which case it is a magnet).

Q3 See page 108 for the magnetic field pattern for a single bar magnet. If the magnet were made stronger, the field pattern would not change, but it is common to indicate the increased field strength by drawing more lines between the original ones.

2 Electrical quantities (page 123)

Q1 a 10.33 A

Use the equation $I = Q/t$, $I = 10/30 = 0.33$ A.

b (i) 10 C; (ii) 36 000 C.

Rearranging the formula, $Q = It$,
(i) $Q = 10 \times 1 = 10$ C, (ii) $Q = 10 \times 60 \times 60 = 36\ 000$ C.

Q2 a 600 joules per second = 600 W

b 50 A

c 2.72 A (or 2.7 A)

d Lower current means that you can use thinner wire; can use simpler switches; will have less energy loss in the wires. But 220 V will be a safety hazard near the water and will require careful protection for the motor and the operator.

Q3 a Try the experiment and see!

b The balloons will acquire the same electrostatic charge, either + or – depending on the cloth used. Either way, like charges will repel each other.

c If the car is stationary then presumably little current is going through the variable resistor. If the car is going at full speed, then the resistor has been taken to zero ohms and the current is not flowing through the resistance wire. In fact, the resistor gets hottest when the p.d. across the resistor is the same as the p.d. across the motor.

3 Electric circuits (page 141)

Q1 a A_1 reading 0.2 A, A_2 reading 0.2 A.

The current is always the same at any point in a series circuit.

b A_4 reading 0.30 A, A_5 reading 0.15 A.

The ammeter A_6 has been placed on one branch of the parallel circuit. Ammeter A_5 is on the other branch. As the lamps are identical the current flowing through them must be the same. Ammeter A_4 gives the current before it 'splits' in half as it flows through the two parallel branches.

c Reading on $V_1 = 6$ V.

The potential difference across the battery is 9 V. This must equal the total p.d. in the circuit. Assuming there is no loss along the copper wiring the p.d. across the lamp must be $9 - 3 = 6$ V.

Q2 a

Inputs			Output
A	**B**	**C**	
0	0	0	I
0	I	I	0
I	0	I	0
I	I	I	0

'Way point' C is the output from an OR gate. This becomes the input for the NOT gate.

b

Inputs			Output
A	**B**	**C**	
0	0	0	I
0	I	0	I
I	0	0	I
I	I	I	0

'Way point' C is the output from an AND gate. This becomes the input for the NOT gate.

Q3

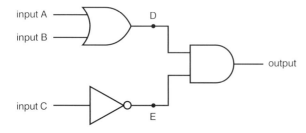

Inputs					Output
A	**B**	**C**	**D**	**E**	
0	0	0	0	1	0
0	0	1	0	0	0
0	1	0	1	1	1
0	1	1	1	0	0
1	0	0	1	1	1
1	0	1	1	0	0
1	1	0	1	1	1
1	1	1	1	0	0

Add 'way points' D and E.
D is A OR B.
E is NOT C.
Output is D AND E.

Q4　**a**　8 V

　　b　1.09 V (or 1.1 V)

　　c　To switch the high voltage and high current of the heater; to control a.c. electricty with a d.c. circuit; to isolate the control circuit from the hazard of mains electricity.

4 Dangers of electricity (page 143)

Q1　**a**　920 W

　　　Power = $V \times I$ = 230 × 4 = 920 W.

　　　Don't forget the unit. Power is measured in watts.

　　b　Water greatly increases the hazard of any contact with mains electricity as the skin becomes highly conducting. If the hairdryer is dropped in the washbasin, then any attempt to pick it out will be fatal.

Q2　You should fit an MCB of 20 A. If the supply were 120 V a.c., you should fit an MCB of 40 A.

5 Electromagnetic effects (page 151)

Q1　Increasing the current flowing through the coil. Increasing the number of coils. Adding a soft iron core inside the cardboard tube.

　　These will produce a stronger magnetic field.

Q2　When the switch is pressed, the electromagnet attracts the hammer support and the hammer hits the bell. The movement of the hammer support breaks the circuit and so the electromagnet ceases to operate. The hammer support then returns to its original position, forming the circuit again and the process is repeated.

　　The electromagnet is constantly activated and then deactivated. This means that the hammer will continually hit the bell and then retract.

Q3　**a**　Iron.

　　　The core must be a magnetic material. It concentrates the magnetic effect.

　　b　A magnetic field is produced.

　　　Remember that a current flowing in a wire will produce a magnetic field around it.

　　c　A current is induced.

　　　The varying magnetic field in the core induces a current in the secondary coil.

　　d　24 V.

　　　$V_p/V_s = N_p/N_s$; $V_p/14 = 12/7$;

　　　$V_p = 12/7 \times 14$ = 24 V.

Q4　If the rotation of the armature is producing electricity which is being withdrawn via A and B, then it is a generator. If, alternatively, electricity is being supplied via A and B to produce rotation then it is a motor.

　　The same device can be used as either a motor or a generator. A motor needs an electrical supply in order to produce movement. In a generator the movement is used to generate electricity.

6 Cathode ray oscilloscopes (page 155)

Q1　Electrons emitted from a heated cathode.

　　The clue is in the name – 'therm' relates to heat and 'ionic' relates to electrically charged particles (electrons in this case).

Q2

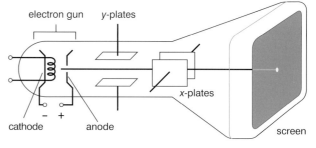

Remember that x-plates are horizontal, y-plates are vertical – just like the axes of a graph!

Q3 **a** 0.8 V

The amplitude is from the centre of the signal to the top (or the bottom). Here it is four squares at 0.2 V per square = 0.8 V.

b 50 ms = 0.05 s

One complete cycle covers five squares horizontally. Each square is worth 10 ms, so 50 ms overall.

c 20 Hz

Frequency = 1/time period = 1/0.05 (The time needs to be in seconds.) Remember the unit for frequency is Hertz (Hz).

ATOMIC PHYSICS

1 Radioactivity (page 165)

Q1 **a** Alpha particles.

b Beta particles.

c Gamma rays.

Alpha particles are the most ionising, gamma rays are the most penetrating.

Q2 **a** A tracer is a radioactive isotope used in detection.

Tracers are widely used to detect leaks and blockages.

b Sodium-24. The lawrencium-257 has too short a half-life; the sulphur-35 and carbon-14 have half-lives which are too long.

The tracer must be radioactive for long enough for it to be detected after injection into the body but must not remain radioactive in the body for longer than necessary.

Q3 **a** Lead.

In fact the uranium-238 decays through a chain of short-lived intermediate elements before forming lead.

b The ratio of uranium-238 to lead-207 enables the age of the rock to be determined.

In a similar way the carbon-14: carbon-12 ratio is important in finding out the age of previously living material.

2 The nuclear atom (page 169)

Q1

Atom	Symbol	Number of protons	Number of neutrons	Number of electrons
Hydrogen	$^{1}_{1}H$	1	0	1
Carbon	$^{12}_{6}C$	6	6	6
Calcium	$^{40}_{20}Ca$	20	20	20
Uranium	$^{238}_{92}U$	92	146	92

Remember:
number of electrons = number of neutrons in a neutral atom
number of neutrons = mass number – proton number

Q2 **a** 11 protons, 13 neutrons.

The atomic number gives the number of protons. The difference between the mass number and the atomic number equals the number of neutrons.

b 15 hours.

The count falls from 100 Bq to 50 Bq in 15 hours. It also falls from 50 Bq to 25 Bq in 15 hours.

Q3 **a** Beta particle.

A beta particle is an electron. The 'beta' symbol can also be written as an electron, 'e'. The electron is shown with an atomic number of –1 and a mass number of 0.

b $x = 24$; $y = 12$.

The mass numbers must balance on the left-hand and right-hand sides of the equation (24 = 24 + 0). The atomic numbers must balance on the left-hand and right-hand sides of the equation (11 = 12 – 1).

ANSWERS TO EXAMINATION QUESTIONS

The University of Cambridge Local Examinations Syndicate bears no responsibility for the example answers to questions taken from its past question papers which are contained in this publication.

GENERAL PHYSICS (pages 50–3)

1 a i Accelerating at a constant rate

    ii Maintaining a constant speed

  b Similarity: the speed is the same (6 m/s)

    Difference: the direction is opposite

  c i $3 \times 20 = 60$ m

    ii $6 \times 52 = 312$ m

2 a i 7(.0) s

    ii Between points P and Q

    iii 22 m

  b i The deceleration is uniform. The deceleration is slower and the bus takes longer and further to stop.

    ii 1.9 m/s^2

  c i The graph shows that the acceleration has a constant magnitude. As the mass is constant, the magnitude of the force must be constant also.

    ii The force is radially inwards, towards the centre of the circular track.

3 a i The law states that the extension of the spring is proportional to the force applied to it.

    ii If the force is doubled from 250 N to 500 N, the extension should double. 2×0.096 m $= 0.192$ m. So this is correct. The same applies if the force is tripled to 750 N or quadrupled to 1000 N.

  b i 84(.0) J

    ii 33.6

4 a They need to know the total height (or the height of each stair and the number of stairs).

    They need the weight (in N) of the student, or the mass (in kg) plus the value of g.

They need to measure the time with a stopwatch. Errors can be reduced by repeating the measurements several times and averaging the results.

  b Work done = force (or weight) $\times$ distance moved against the force

  c i Potential energy has increased.

    ii She has generated kinetic energy to run up the stairs and to move her arms, she has created heat energy.

5 a 500 000 Pa, 500 000 N/m^2, 500 kPa or 500 kN/m^2 are all equivalent answers.

  b 5250 N

6 a From B to D the speed increases at a constant rate (straight line at an angle upwards from B).

    From D to E the speed is constant (horizontal line above the x-axis and parallel to it).

    From E to F the speed decreases quickly back to zero. (Straight line back to x-axis with a steeper gradient than BD.)

  b i 1.2 J

    ii 0.625 J

  c The speed is a scalar and only has a magnitude. This magnitude is unchanged. The velocity has a magnitude and a direction. The direction of movement changes, and so the velocity has changed.

  d The block has been slowed between E and F by friction, and this will have generated heat energy. So the initial potential energy equals the final potential energy plus the heat energy.

THERMAL PHYSICS (pages 82–3)

1 a i Constantan or any metal other than copper will work.

ii Z is a galvanometer, milliammeter or a millivoltmeter.

b You have the two junctions at different temperatures, one hot and one cold. The temperature difference produces a small e.m.f. that causes a current to flow through the meter. One junction is kept at a known temperature (often the melting point of pure ice). The meter must be calibrated against other known temperatures.

c A dull black surface will be hottest.

2 a 21 mm

b i You put the cylinder of the apparatus in a beaker of melting ice, and in the steam over boiling water. In each case you wait for movement to stop. You mark the position of the piston at the two temperatures, and divide the length between into a scale from 0 to 100.

ii You immerse the cylinder in water at least as far as the piston, and you wait for the piston to cease moving. You then use the scale to read the temperature.

3 a Some of the water molecules have more energy than most. At any temperature some molecules will have sufficient velocity to escape from the surface of the water.

b Evaporation occurs only at the surface of the liquid, while boiling occurs throughout the liquid. Evaporation occurs to some extent at any temperature; boiling occurs only at the boiling point, which is fixed for a given liquid and a given atmospheric pressure.

c 2250 J/g

PROPERTIES OF WAVES, INCLUDING LIGHT AND SOUND (pages 104–5)

1 a The drawing should show three more circular arcs moving out from the gap. The spacing between waves should be unchanged.

b 6×10^{-7} m

2 a i The C should be marked on any peak, or on the axis below the peak. The R should be marked on an adjacent trough or on the axis above the trough.

ii The distance between C and R is half a wavelength.

b 260 Hz

3 a The letters C and R should alternate along the line and be equi-spaced.

b i A compression is a region of high pressure with the air molecules closer together than average.

ii A rarefaction is a region of low pressure with the air molecules further apart than average.

c In a longitudinal wave the movement of the particles is to and fro in the same direction that the wave is travelling.

d 0.29 s (or 0.3 s)

4 a i The incident ray is in air and at 90° to the wavefronts; the refracted ray is in the glass at 90° to the wavefronts in the glass; the normal is drawn from the air and into the glass, at 90° to the edge of the glass. The incident ray and the refracted ray both terminate on the edge of the glass, and all three lines should meet at the same point on the edge of the glass.

ii The angle of incidence is the angle between the incident ray and the normal; the angle of refraction is the angle between the refracted ray and the normal.

iii Angle of incidence 45°, angle of refraction 30°.

b Refractive index = 1.41

ELECTRICITY AND MAGNETISM (pages 156–9)

1 a Some of the electrons in the atoms that make up one sphere are rubbed off and onto the other sphere. The sphere with too few electrons becomes positively charged and the one with too many becomes negative.

b The lines start by going directly away from the + sphere at 90°. They then bend round and move over to the − sphere. The last part of the lines approach the − sphere at 90°. Each line should have an arrow pointing along the line towards the − sphere.

c i 3.2×10^{-9} C (3.2 nC)

ii 3.2×10^{-6} A (3.2 μA)

2 a 1.5A

b i 10 Ohms

ii 2 Ohms

c 72 W

d i 12 V

ii 6 V

e i less resistance

ii less resistance

3 a i A transistor may be used as a switch, as a form of relay, or as an amplifier of a small signal.

ii As a switch, the circuit through the collector and emitter switches on if the p.d. between the base and emitter exceeds 0.6 V, and switches off otherwise. You can vary the p.d. between base and emitter either by connecting the base to a potential divider, or by feeding a small current into the base.

As an amplifier, small changes in the current into the base produce large changes in the current through the collector and emitter.

b i

```
input ─┐
        ⊐D─ output
input ─┘
```

ii If either input is on, then the output is on. If both inputs are on then the output is on, but if both inputs are off then the output is off. ("High" and "1" are synonyms for ON and "low" and "0" are synonyms for OFF.)

4 a i The signal is regular, and constantly repeats the same pattern, but it is not a sine wave. It consists of several waves added together.

ii 1.6 V

iii You connect a known voltage across the Y plates, for example a 1.5 V cell, and measure the displacement of the horizontal line on the screen.

b i 6.1 cm

ii 31 ms

iii There are many examples of events that are separated by small time intervals. Runners finishing a race, car breaking two light beams across the road etc.

5 a 1520 W (or 1.52 kW)

b i In parallel means that each appliance is connected across the mains supply.

ii The appliances can be switched on individually; a fault in one appliance does not affect the others; the voltage applied to each appliance is a constant, no matter what else is switched on.

c i 0.83 A (0.8 A)

ii 12 960 000 J (12.96 MJ)

iii 960 Ohms

6 a i The field line is a circle, centred on the wire, and passing through P. The arrow on the wire should point anticlockwise.

ii The arrow through Q should point to the left.

b i If the current is increased the compass will either not move at all or will move negligibly.

ii If the current is reversed, the compass will point the other way.

c i At S the field is stronger.

ii At T the field is the same.

iii At W the field is the same.

ATOMIC PHYSICS (pages 170–1)

1 a i The apparatus must consist of a source of radiation in a container with an aperture for the radiation to emerge; a detector (such as a Geiger counter) and suitable absorbers, which can consist of material to place between the source and the detector. Air is also an absorber and so the air gap can be labelled for this purpose.

ii You measure the background rate on the detector before the source is place in front of it (or with a thick lead shield in front of the source.) You then take readings with the source exposed to the detector, and with various absorbers placed between the detector and the source.

iii If the radiation is all absorbed by 10 cm of air, paper or cardboard, then it is alpha.

b The path will be curved at right-angles to the magnetic field, and will bend away from the observer, towards the paper.

2 a i $^{24}_{11}\text{Na} \rightarrow {}^{24}_{12}\text{Mg} + {}^{0}_{-1}\beta$

ii The path will be curved at right-angles to the magnetic field, and will bend towards the observer, away from the paper.

b i A detector such as a geiger counter is used to detect the radiation coming out of the body. It can either be held a fixed distance from the injection to measure how fast the blood is flowing, or it can be scanned across the body to see which areas have the greatest flow.

ii Alpha particles are unable to escape through the skin, and beta particles will largely be lost while they escape from the body. Most of the gamma rays will travel from the radioisotope to the detector.

GLOSSARY

air resistance

amplitude

background radiation

centripetal force

compression

acceleration How much something's speed increases every second. Acceleration = change in speed / time taken to change.

air resistance Drag caused by something moving through air.

alternating current (a.c.) Electrical current that repeatedly reverses its direction, like mains electricity.

ammeter Instrument that measures electrical current in amperes.

ampere Unit of current. The electric charge that flows during one second.

amplitude Maximum charge of the medium from normal in a wave. For example, the height of a water wave above the level of calm water.

analogue Describes a quantity that can charge smoothly, like the position of a pointer over a dial (the opposite of digital).

background radiation The level of radiation found due to natural processes in the environment.

centripetal force Force that acts towards a centre. A centripetal force is needed to move in a circle.

charge Fundamental property of matter that produces all electrical effects.

compression A region where particles are squashed together.

conductors, heat Substances that conduct heat very well.

conventional current Movement of positive charge that is imagined to move from the positive terminal to the negative terminal of a battery. Equivalent in effect to the real flow of negative charge in the opposite direction.

coulomb Unit of electric charge.

crest The highest part of a wave.

density The mass, in kilograms, of one metre cube of a substance.

diffraction Waves spreading into the shadow when they pass an edge.

diffusion Molecules moving from an area of high concentration to an area of low concentration.

digital Describes quantities that can only be displayed as numbers (the opposite of analogue).

diode Device that only lets electricity flow through it one way.

direct current (d.c.) Current that always flows in the same direction.

dispersion Splitting white light into colours.

efficiency Ratio of the useful work done by a machine to the energy input, often written as a percentage.

elastic Describes material that go back to their original shape and size after you stretch them.

electric current Flowing electric charge.

electric field Region in which any electrical charges will feel a force.

electromagnetic induction A changing magnetic force can induce electric current in a wire.

electromagnetic spectrum The 'family' of electromagnetic radiations (radio, microwave, infrared, visible light, ultraviolet, X-rays, gamma rays). They all travel at the same speed in a vacuum.

electromagnets Magnets made from a coil of wire. The magnetic force is made when electric current flows in the coil. The magnetic force is stronger when the coil is wrapped around a piece of iron.

electron Negatively charged particles with a negligible mass that form the outer portion of all atoms.

extension The increase in length when something is stretched.

fluid Any liquid or gas.

force A push or a pull, measured in newtons (N).

free fall Movement under the effect of the force of gravity alone.

frequency The number of vibrations per second, measured in hertz (Hz).

friction The force that resists when you try to move something.

fuse A special wire that protects an electric circuit. If the current gets too large, the fuse melts and stops the current.

gamma rays See electromagnetic spectrum.

gradient Slope of a curve.

gravitational field strength The force of gravity on a mass of one kilogram. The unit is the newton per kilogram.

gravitational potential energy Objects have more potential energy when they are higher up in the Earth's gravitational field.

half-life Time it takes for half of a sample of radioactive nuclei to decay.

induce To affect something without touching it. An electric force can induce charge in a conductor. A changing magnetic force can induce electric current in a wire.

insulators of heat Substances that do not conduct heat very well.

interference Waves combine with each other as they collide.

inversely proportional Two quantities are inversely proportional if one doubles when the other halves.

elastic

electric current

electric field

electromagnetic spectrum

free fall

friction

kinetic energy

magnetic materials

mass

parallel

photons

power

ionising radiation Charged particles or high-energy light rays that ionise the material they travel through.

joules The unit of energy. One joule is the energy needed to push an object through one metre with a one newton force.

kinetic energy Moving objects have kinetic energy. Fast, massive objects have more kinetic energy than slow light objects.

law of energy, first Energy cannot be created or destroyed.

law of energy, second Some energy always becomes unusable whenever energy is transferred.

longitudinal wave Wave where the change of the medium is parallel to the direction of the wave.

magnetic field Region in which magnetic materials feel a force.

magnetic materials Materials that are attracted to magnets and can be made into magnets. Iron, cobalt and nickel are magnetic materials.

mass Amount of material in an object, measured in kilograms.

momentum Fast objects and massive objects need a lot of force to stop them – they have a large momentum. Momentum = mass x velocity.

neutron Particle present in the nucleus of atoms that has mass but no charge.

nucleus, atomic The tiny centre of an atom made from protons and neutrons.

Ohm's law The current flowing through a component is proportional to the potential difference between its ends, providing temperature is constant.

parallel Describes a circuit in which the current splits up into more than one path.

photons Particles of light and other electromagnetic radiations. Sometimes radiation behaves like waves, sometimes like particles.

pitch Whether a note sounds high or low to your ear.

polarity Some components only work correctly when connected the right way around – with the right polarity.

potential difference (p.d.) The energy transferred from one coulomb of charge between two points. Measured in volts. Often called the 'voltage'.

power Amount of energy transferred every second. The energy can be transferred *from* somewhere (e.g. a power station) or *to* somewhere (e.g. an electric kettle). Power = energy transferred / time taken.

pressure The effect of a force spread out over an area.
Pressure = force / area.

primary coil The input coil of a transformer. You connect it to the voltage you want to change.

proton Positively charged, massive particles found in the nucleus of an atom.

radiation Energy that travels in straight lines, e.g. electromagnetic rays.

radioactive Describes a substance that has nuclei that are not stable.

radioactive decay Natural and random change of a nucleus.

rarefaction Region where particles are stretched further apart than normal.

reflection When waves bounce off a mirror. The angle of incidence is the same size as the angle of reflection.

refraction When waves change direction because they have passed into a different medium. They change direction because their speed changes.

resistance Property of an electrical conductor that limits how easily an electric current flows through it. Measured in ohms.

resultant force A single imaginary force that is equivalent to all the forces acting on an object.

series Describes a circuit in which the current travels along one path through every component.

short-circuit Unwanted branch of an electrical circuit that bypasses other parts of the circuit and causes a large current to flow.

spectrum The 'rainbow' of colours that make up white light (red, orange, yellow, green, blue, indigo and violet).

speed How far something moves every second.
Average speed = distance travelled / time taken.

transformer Machine that changes the voltage of a.c. electricity.

transverse wave A wave where the change of the medium is at 90° to the direction of the wave.

trough Lowest part of a wave.

variable resistor Component with a resistance that can be manually altered.

velocity Quantity that indicates the speed and direction of an object.

volt Unit of voltage. Energy carried by one coulomb of electric charge.

watt Unit of power. One watt is one joule transferred every second.

wavelength The distance between the same points of successive waves. For example, the distance from one crest to the next.

weight Force of gravity on a mass. The unit of weight is the newton.

work Energy transferred when a job is done.
Work = force x distance moved in direction of force.

reflection

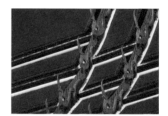

resultant force

series

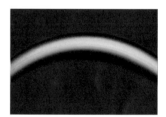

spectrum

transformer

weight

INDEX